Knowledge Sharing on the Shopfloor

Sven Hoffmann

Knowledge Sharing on the Shopfloor

Exploring the Use of Augmented Reality in Industrial Set-up Practices

Springer Vieweg

Sven Hoffmann
FernUniversität in Hagen
Hagen, Germany

FernUniversität in Hagen, Fakultät für Mathematik und Informatik, Hagen Deutschland
Dissertation zur Erlangung des Grades eines Doktor-Ingenieurs (Dr.-Ing.)
1. Gutachter: Prof. Dr. Thomas Ludwig (FernUniversität in Hagen)
2. Gutachter: Prof. Dr. Volker Wulf (Universität Siegen)
3. Gutachter: Prof. Dr.-Ing. Jörg M. Haake (FernUniversität in Hagen)
Tag der Disputation: 13.05.2025

ISBN 978-3-658-50132-7 ISBN 978-3-658-50133-4 (eBook)
https://doi.org/10.1007/978-3-658-50133-4

This Springer Vieweg imprint is published by the registered company Springer Fachmedien Wiesbaden GmbH, part of Springer Nature.
The registered company address is: Abraham-Lincoln-Str. 46, 65189 Wiesbaden, Germany

Competing Interests The author has no competing interests to declare that are relevant to the content of this manuscript.

Abstract

Technological progress has made significant advances in manufacturing companies during the so-called fourth industrial revolution (Schwab, 2017). Production machines and their operation—combined with ever smaller batch sizes—are leading to a sharp increase in the demands placed on personnel in production environments (Ludwig et al., 2018). The ability to master knowledge-intensive production processes characterises manufacturing companies and holds existential relevance (Davis et al., 2012). As a result, companies have a strong demand for qualified personnel, who are increasingly leaving companies due to demographic changes (Lisiankova & Wright, 2005). Technological developments in the course of the fourth industrial revolution have led to machines and systems having a higher degree of automation (Brynjolfsson & McAfee, 2014). To a certain extent, set-up processes on machines and systems are an exception, as these are generally characterized by a high level of manual effort and complexity, which means that they are knowledge-intensive activities (Goubergen & Lockhart, 2005). Therefore, there is a strong need for research into knowledge- and expertise-sharing practices in companies and their accompanying technical assistance, which has a long tradition in the academic field of computer-supported cooperative work (Ackerman et al., 2013).

This dissertation addresses these discussions by exploring manual set-up processes on production machines, seeking to provide insight into how cyber-physically embedded augmented reality (AR) technologies should be designed to support knowledge and expertise sharing processes. Based on design case studies (Wulf et al., 2015) in manufacturing companies, the main design implications are first derived from a practice-oriented perspective and then translated into implementations of AR technologies. The practical evaluation of AR tools provides insights into how AR technologies support knowledge on the shopfloor.

This work's results characterise set-up processes by highlighting the nuances of knowledge sharing in the course of set-up activities. The AR applications provide important results regarding designing holograms to represent tools, machines, and operator interfaces. The findings illustrate how contextualised in-situ AR instructions strengthen the role of qualified labour in set-up processes, leading to the operator gaining a better understanding of the process.

Related Publications

Parts of this dissertation have already been published as conference or journal papers.

- Chapter 4: De Carvalho, A. F. P., Hoffmann, S., Abele, D., Schweitzer, M., Tolmie, P., Randall, D., & Wulf, V. (2018). Of embodied action and sensors: Knowledge and expertise sharing in industrial set-up. Journal of Computer Supported Cooperative Work (JCSCW), 27, 875–916.
- Chapter 5: Hoffmann, S., de Carvalho, A. F. P., Abele, D., Schweitzer, M., Tolmie, P., & Wulf, V. (2019). Cyber-physical systems for knowledge and expertise sharing in manufacturing contexts: Towards a model enabling design. Journal of Computer Supported Cooperative Work (JCSCW), 28, 469–509.
- Chapter 6: De Carvalho, A. F. P., Hoffmann, S., Abele, D., Schweitzer, M., & Wulf, V. (2021). Designing Cyber-Physical Production Systems for Industrial Set-Up: A Practice-Centred Approach. In IFIP Conference on Human-Computer Interaction (INTERACT) (pp. 678–701). Cham: Springer International Publishing.
- Chapter 7: Hoffmann, S., de Carvalho, A. F. P., Schweitzer, M., Abele, N. D., & Wulf, V. (2022). Producing and Consuming Instructional Material in Manufacturing Contexts: Evaluation of an AR-based Cyber-Physical Production System for Supporting Knowledge and Expertise Sharing. Proceedings of the ACM on Human-Computer Interaction, 6(CSCW2), 1–36.
- Chapter 8: Jasche, F., Hoffmann, S., Ludwig, T., & Wulf, V. (2021). Comparison of different types of augmented reality visualizations for instructions.

In Proceedings of the 2021 CHI Conference on Human Factors in Computing Systems (CHI 2021), (pp. 1–13), ACM Press New York.
- Chapter 9: Hoffmann, S., Ludwig, T., Jasche, F., Wulf, V., & Randall, D. (2023). RetrofittAR: Supporting Hardware-Centered Expertise Sharing in Manufacturing Settings through Augmented Reality. Journal of Computer Supported Cooperative Work (JCSCW), 32(1), 93–139.

Contents

Part II Findings

Abbreviations

AAR	Abstract AR
AAR+V	Abstract AR with Videos
AG	Answer Garden
AG2	Answer Garden 2
AR	Augmented Reality
CAD	Computer-Aided Design
CAM	Computer Aided Manufacturing
CAR	Concrete AR
CAR+V	Concrete AR with Videos
CHI	ACM Conference on Human Factors in Computing Systems
CIM	Computer Integrated Manufacturing
CLT	Cognitive Load Theory
CNC	Computer Numerically Controlled
CPPS	Cyber–Physical Production System
CPS	Cyber–Physical System
CSCW	Computer-Supported Cooperative Work
DCS	Design Case Study
DW	Design Workshop
FEM	Finite Element Method
GD	Grounded Design
GUI	Graphical User Interface
HCI	Human Computer Interaction
HHD	Hand-Held Display
HMD	Head-Mounted Display

HMI	Human–Machine Interface
ICT	Information and Communication Technology
INTERACT	IFIP Conference on Human-Computer Interaction
IoT	Internet of Things
IS	Information Systems
JCSCW	The Journal of Collaborative Computing and Work Practices
KES	Knowledge and Expertise Sharing
KM	Knowledge Management
M	Mean
MRTK	Mixed Reality Toolkit
N	Number of Participants
OM	Organisational Management
OPC-UA	OPC Unified Architecture
P	Participant
PD	Participatory Design
R	Radius
R&D	Research and Development
RSME	Rating Scale of Mental Effort
SCM	Supply Chain Management
SD	Standard Deviation
SDK	Software Development Kit
SECI	Socialisation, Externalisation, Combination and Internalisation
SME	Small and Medium Enterprise/Small and Mid-sized Enterprise
SMED	Single-Minute Exchange of Die
SUS	System Usability Scale
TA	Thematic Analysis
UEQ	User Experience Questionnaire
UI	User Interface

List of Figures

List of Tables

Introduction 1

1.1 Motivation

The complexity of modern machines and thus hardware-related activities has increased immensely in recent years. The drivers of complexity can be attributed to technological advances in the course of the so-called fourth industrial revolution and the market requirements regarding the fulfilment of mass customisation capabilities (Müller et al., 2018; Schwab, 2017; Ullah & Narain, 2021). This poses significant challenges for manufacturing companies in mastering value-adding production processes, given that—in addition to the technical advances in production equipment—the workforce must be further developed to ensure that the processes and equipment are mastered by production personnel in day-to-day production. Specifically, this will result in new jobs with new job profiles and a higher level of technical expertise will be required. It is therefore imperative to combine technical advances with measures to support employees (Müller et al., 2018). These developments lead to the emergence of knowledge-embedded facilities. Within these facilities, the workforce will hold fundamental importance and significantly contribute to competitive advantage (Davis et al., 2012). Taking a closer look at this macro perspective regarding different job profiles, technological progress will lead to job profiles becoming more professionalised due to higher qualification requirements. The path to higher qualifications must be supported by socio-technical systems on the shopfloor (Hirsch-Kreinsen, 2016).

The need for socio-technical support for knowledge transfer becomes even greater if we add a demographic perspective to the need for a more highly qualified workforce. The EU population is predicted to decline in the 20 to 64 age group, which will have direct implications for the labour market as the potential labour supply will also experience a decline (Lisiankova & Wright, 2005). The

S. Hoffmann, *Knowledge Sharing on the Shopfloor*,
https://doi.org/10.1007/978-3-658-50133-4_1

decline in the workforce from the baby boomer generation is particularly severe, especially as a wealth of knowledge and expertise is also disappearing from companies with this generation. For this reason, companies must be able to transfer the knowledge and experience of the older generations (Calo, 2008).

The combination of increasing complexity in the operation of machines and demographic challenges means that researching new human-machine interfaces is one possibility at the operational level of the machine operator, which can lead to a fusion between the machine, process and human. The interface design is a socio-technical issue that must be based on the premise of designing an expertise sharing support system providing the best possible support for the worker in complex situations, capturing the production context, and integrating the company's knowledge content (Gorecky et al., 2014). The demands on hardware and software design are high as the technology must be integrated into the day-to-day work of the knowledge carriers and those in need of knowledge on the shopfloor. This is where the use of augmented reality (AR) technologies in the form of a head-mounted display (HMD) comes into play. The hardware offers the possibility of interaction during mechanical activities—for example, at a machine—and AR representations can guide knowledge-intensive processes (Feiner et al., 1993). In summary, the question emerges concerning designing AR applications to support knowledge management in hardware-related manual activities on production machines. This work draws on the application field of set-up processes on industrial machines.

Set-up processes comprise preparatory steps on a machine or tool before production starts (Palanisamy & Siddiqui, 2007). The separation between set-up steps and subsequent production is particularly fluid when manufacturing small batch sizes, as constant adjustment activities characterise the operator's work. This does not necessarily have to involve mechanical settings on the machine, although the production parameter settings on the machine are also preparatory activities and therefore part of the set-up process. Modern production machines have many setting parameters, increasing the complexity of the set-up processes. A successful set-up process is the basic prerequisite for starting a production process. Given that batch sizes and delivery sizes are declining due to the increasing flexibility of the world market and the increasing customer orientation, more set-up processes are taking place (Van Goubergen & Van Landeghem, 2002). Therefore, the priority is to support set-up processes and make them as economical as possible. Since set-up processes are mostly manual tasks carried out by trained employees, people with experience and knowledge play a decisive role in the execution of set-up processes. The challenge here is to reduce the use of resources while meeting the demand for increasingly flexible production cycles and diverse quality requirements (Shingo, 1985). Depending on the machine to be set up, set-up processes can be individual activities by one person or collaborative activities by several people. These developments also have a direct impact on employees, exerting an increased demand for flexible skills to handle heterogeneous set-up operations. This means that machine setters must be able to produce a wide range of products with various materials and different tolerance levels on different machines. Combined with a predominantly manual operation, it is clear that set-up knowledge significantly influences the set-up process's success but has received little or no attention to date (Goubergen & Lockhart, 2005). It is therefore important to research how modern technologies can support exchanging knowledge and experience in set-up processes.

Knowledge and expertise sharing have a long history in computer-supported cooperative work (CSCW) research, not least because knowledge and expertise sharing (KES) helps companies to position themselves competitively and develop in line with the market (Bolisani & Scarso, 1999; Spender, 1996a). To date, two major generations have been identified in the CSCW literature to consider KES (Ackerman et al., 2013). However, embedding expertise sharing in practice is a major challenge, accompanied by the well-known problem of how to capture, describe, and demonstrate knowledge to be assimilated by others (Nonaka & Takeuchi, 1995; Schmidt, 2012). Given that the notation of tacit knowledge mystifies knowledge management and distracts from the actual problems of knowledge management, there are reservations about this notation (Schmidt, 2012). Consequently, it is not necessary to consider the theoretical foundations but rather the practical manifestations of tacit knowledge work. This thesis shows that so-called "tacit" knowledge in practice conceals several possibilities to articulate knowledge in different ways, although strong contextual and local influences are present (Randall et al., 1996). A similar argument concerns know-how (Ryle, 1945, 2009), characterised by the absence of propositional knowledge components. Propositional knowledge components do not necessarily have to be converted into non-propositional knowledge components, although modern technologies can be used to convey knowledge content without using formalisms of externalisation. It is important to note that the different types of possible technical support are configured to reflect the types of knowledge and expertise and the different modes through which they are displayed. This is where modern technologies in the form of AR-based cyber-physical systems (CPS) come in, as these technologies have the potential to enable new ways of capturing knowledge and reproducing knowledge (Feiner et al., 1993).

In the CSCW discourse, a new third generation of expertise sharing is dawning. The technology is seen as having a multitude of possibilities, primarily arising from the networking of different systems and sensors to connect the cybernetic and physical worlds (Lee & Seshia, 2015; Monostori et al., 2016; Rajkumar et al., 2010). Challenges arise at two levels: at the technical level, connectivity needs to be ensured (Pacaux-Lemoine et al., 2018; Wang et al., 2015), while at the interaction level, research should explore how users deal with this new technology in the course of supporting knowledge transfer on the shop floor. In any case, there has been no further research on how AR-based CPS technologies can be embedded in work practices on the shop floor, especially in set-up practices, and how these technologies must be designed to support the knowledge-transfer process between experts and novices. For this purpose, it is first necessary to examine the knowledge-transfer process in detail and the individual actors, as well as recording the organisational boundary conditions. Building on this, it is necessary to implement and evaluate suitable technologies based on the set-up practices.

1.2 Research Goals and Central Research Question

Van Goubergen and Lockhart (2005) argue that knowledge is a key success factor in implementing a set-up process. Work practices have always been studied in CSCW and knowledge-intensive practices in particular have always been the focus of consideration. In this tradition, two major generations have emerged in considering KES (Ackerman et al., 2013). Modern technology offers new possibilities for supporting knowledge-intensive processes. Based on initial findings regarding the possibilities of CPS and AR technologies (Ludwig et al., 2017; Robertson & Wagner, 2015), this thesis identifies potentials of how the technologies can be used in knowledge- and expertise-sharing processes. The work aims to accurately identify the set-up process and existing challenges, placing a particular focus on the practices of the workers on the shop floor. This approach has been largely under-represented in research to date, while the need for such an approach is clearly given by the growing challenges before and during the production processes. These are not only practices based on KES processes but also social practices between the people involved in the KES processes. Based on the findings regarding KES practices, the thesis explores the technical possibilities to support the identified practices. Accordingly, this thesis addresses and strongly concretises the shortcomings of the literature in identifying technologies and their technical design to support KES practices, based on the example of set-up processes. Modern CPS technologies in particular represent promising approaches, although have yet to be further detailed in terms of human-technology interaction (Pacaux-Lemoine et al., 2018). Therefore, this work identifies opportunities for AR-based CPS technologies to be used in knowledge-intensive set-up processes. The thesis combines the findings from the empirical study with implications for technology design using a knowledge-transfer model. For this purpose, the essential mechanisms and components of knowledge transfer are presented.

This thesis explores set-up processes in the context of a design case study. The empirical requirements assessment focuses on the working practices during the set-up processes, while the traditional understanding of set-up processes is expanded by essential aspects by focusing on knowledge-intensive practices. The implementation based on the design implications includes two AR applications, which are evaluated in practice. The following central question addresses the research objective of investigating and supporting knowledge-intensive set-up processes with the help of AR-based cyber-physical production system (CPPS) technologies:

How can knowledge-intensive set-up processes be supported with the help of modern AR-based CPPS technologies?

1.3 Structure of the Thesis

The thesis consists of three main parts. Chapters 4, 5, 6, 7, 8 and 9, which form the core of the thesis, have already been published in peer-reviewed publications. In this thesis the camera-ready versions of the papers have been used and have been adapted to the format of this thesis with minor modifications (mainly by harmonizing the labelling of images, tables and references).

- **Chapters 1, 2 and 3** motivate the thesis and place the work in the state of the art. For this purpose, the related work is summarised. Furthermore, it presents the research approach and methodology of this work.
- **Chapters 4, 5, 6, 7, 8 and 9** form the core of the thesis. For this purpose, the individual peer-reviewed articles are presented in a thematically sequential collection. The collection starts with empirical results and the design implications derived from them and ends with two implementations and the corresponding evaluations.
- **Chapter 10** summarises the results and discusses the contribution of the thesis against the background of the literature. In addition, an outlook on future research approaches is given.

Part I
Overview

2 Related Work

This chapter presents the results of research in the area of knowledge- and expertise sharing relevant to this thesis from the perspective of CSCW and other relevant research directions (Sect. 2.1). In addition, the current state of the art of supporting knowledge management with the help of new technologies is discussed (Sect. 2.2). Specifically, CPS and AR technologies are examined in further detail. The work focuses on knowledge-intensive areas in production environments, particularly set-up processes. Finally, at the end of the chapter, the research gap is derived and formulated with five specific research questions (Sect. 2.3).

2.1 Knowledge and Expertise Sharing

In CSCW, KES has been a recurrent topic of interest since the late 1980s, when terms such as organisational memory (OM), organisational remembering, collective intelligence, and expertise location became commonplace (Ackerman et al., 2013; Randall et al., 1996). From the outset, a focus was placed on developing systems to support creating an OM that could be accessed by whoever needed it and whenever they needed it (Ackerman, 1998). Ackerman's (1998) work advocates acquiring a detailed empirically-based understanding of the practices associated with knowledge transfer, which had previously been neglected in the literature (Ackerman & Halverson, 2004).

The first generation of KES research in CSCW focused on a repository model to support creating an OM that anyone could access when needed (Ackerman et al., 2013), whose limitations soon became apparent. Among other things, issues regarding the dichotomy between experts and users, problems with the contextualisation of knowledge stored in the system, and the authoring burden stemming

S. Hoffmann, *Knowledge Sharing on the Shopfloor*,
https://doi.org/10.1007/978-3-658-50133-4_2

from such systems emerged (Ackerman & McDonald, 1996). Furthermore, Randall et al. (1996) highlighted that the metaphor of an OM failed to capture the socially situated nuances of processes associated with KES. Hence, attention was drawn to the fact that while OM implies a particular and static conceptualisation of 'skills', 'knowledge', and 'memory', understanding the social context in which the process of 'remembering' (as in accessing information and making use of it in some way) occurs is key to grasping KES-related activities. This change in focus originated in a second generation of KES research in CSCW, mainly focusing on how to support interactions related to sharing knowledge and expertise and the vehicles whereby knowledge is shared among knowledgeable actors (Ackerman et al., 2013).

A further distinction has been made between two research traditions regarding how knowledge and expertise flows among knowledgeable actors, namely 'knowledge sharing' and 'expertise sharing'. These two ideas share many features and overlap but are differentiated by the knowledge externalisation status. While externalisation plays an important role in knowledge sharing, expertise sharing is taken to be less dependent on prior externalisation (Ackerman et al., 2013; Pipek et al., 2012). This resonates with the view that expertise largely corresponds to embodied knowledge (Fitzpatrick, 2003). However, none of this quite eliminates the controversy surrounding 'tacit' and 'explicit' knowledge (Nonaka et al., 2000; Polanyi, 1967).

A core criticism here relates to how the concept of tacit knowledge confuses something that was originally a clear practice-oriented consideration. Schmidt (2012) presents a very strong argument against the dichotomy between tacit and explicit knowledge, expressing that the notion of tacit knowledge mystifies the very concept of practical knowledge. He favours a focus on practices concerning the actual application of knowledge and their concomitant outcomes. For him, the important issue is to a certain who uses which tools and to what level of detail. He also suggests the need for a deeper understanding of 'didactic practices' or 'mutual learning'.

In response to the growing interest in tacit knowledge in the organisational management and information systems literature a debate has also developed in the CSCW literature over several years concerning the distinctiveness of the practical skills and knowledge that people draw upon when undertaking certain tasks (Ackerman et al., 2013; Schmidt, 2012). The acquisition and realisation of practical tasks that are resistant to any propositional formulation of the knowledge exercised in their accomplishment has been described in the literature as 'know-how' (Ryle, 1945, 2009). Meanwhile, an understanding of tasks that can be discussed in a propositional form—often in connection with 'know-why' (Lapre et al., 2000)—has been described as 'know-that' (Ryle, 1945, 2009). This discussion is important concerning production processes as it influences how various production activities should be performed and presented, as well as how this information can be shared with others who wish to perform similar activities.

Another distinction in the sharing of knowledge and expertise in CSCW research relates to the rhetorical focus. According to Ackerman et al. (2013), the first generation of research focused on the technical aspects of knowledge management, strongly emphasising creating databases of instructions and standard operating procedures. However, neglecting the lack of incentive structures (Hsieh et al., 2011; Orlikowski, 1993; Tausczik & Pennebaker, 2012) and the social-technical gap led to new problems.

Once this socio-technical divide was recognised, CSCW research increasingly focused on examining the organisational and social aspects of knowledge and experience sharing, which was seen as critical to the success of knowledge management initiatives (Ackerman & Halverson, 2004), particularly for local and personal knowledge that is difficult to capture (Ackerman et al., 2003). This has led to a stronger emphasis on concepts such as communities of practice (CoP) (Ackerman et al., 2013; Wenger, 1998) and social capital (Nahapiet & Goshal, 1998). CoPs offer insights into the relationship between shared practices and learning opportunities, while social capital helps to understand the conditions that can promote sharing knowledge and expertise.

Numerous studies from different fields have contributed to the field of KES. One notable work in this context is the study by Randall et al. (1996) at the organisation level, highlighting the crucial role of a specific person in the transfer of 'local knowledge'. Another important study was conducted at Xerox PARC on information sharing between service technicians (Yamauchi et al., 2003), based on an earlier study by Orr (1996). This study analysed how service technicians supported each other with the relevant details to deal with printer repair exceptions. Some work also explores knowledge sharing within and between different scientific communities (Velden, 2013), emphasising the influence of personal interests. In addition, studies have investigated how documents are shared and reused, as a key feature of knowledge work (Harper et al., 2001).

2.1.1 Organisational Memory

Ackerman (1998) has dealt intensively with the concept of OM, focusing on storing organisational knowledge in artefacts such as relational databases (Ackerman, 1998; Randall et al., 1996). An essential step in OM is the documentation of the organisational structure to capture features that might be relevant for "memory" (Ackerman & Malone, 1990). The relevant literature discusses potential barriers to storing and utilising OM. First, accessing memory can be cumbersome (Randall et al., 1996), and numerous problems can arise, especially in distributed organisations or when many people are involved (Ackerman & McDonald, 1996).

However, OM involves more than simply data storage. Various systems and approaches have emerged over time, although there has never been a definitive agreement on what kind of support such systems should provide (Bannon & Kuutti, 1996). At a deeper level, the metaphor of 'memory' has also been questioned. For example, Randall et al. (1996) argue that essential social aspects are not considered and only a few examples of these aspects are included. This critique strongly resonates with discussions on the grammatical order of concepts such as 'knowing how' and 'knowing what' outlined above.

2.1.2 Other Relevant Studies of Knowledge and Expertise Sharing

In recent decades, several fields have taken an interest in KES in organisational contexts. While researchers in CSCW have sought to understand and support the process of sharing, researchers in fields such as OM and IS have focused on knowledge creation and management. Much of the work here draws upon Polanyi's concept of tacit knowledge (Nonaka, 1988; Nonaka et al., 2000; Nonaka & Takeuchi, 1995; Nonaka & Von Krogh, 2009). In these studies, organisations are seen as both information-processing machines and entities that create knowledge through the action and interaction of their components (Nonaka et al., 2000). Understanding the essence of this process is argued to be key to successful knowledge management and transfer. To tackle this, Nonaka et al. (2000) set out to develop a knowledge-based theory of the firm.

The SECI model—as termed by the authors—is a spiral-based model that covers four processes of 'knowledge conversion', namely socialisation, externalisation, combination, and internalisation (Nonaka et al., 2000). Socialisation is the process through which knowledge sharing is negotiated between individuals, usually through direct social interaction, observation, and imitation. Externalisation describes the process whereby actors attempt to transform implicit into explicit knowledge through (linguistic) articulation. At the point of combination, explicit knowledge elements are exchanged, linked, and enriched at different levels. Here, explicit knowledge—often articulated through documents or electronic communication—is exchanged between individuals. During internalisation, knowledge seekers attempt to internalise explicit knowledge (Nonaka et al., 2000). This is a difficult yet important aspect of creating new knowledge, only arising when existing (experiential) knowledge has been internalised through socialisation, externalisation, and combination (Döring, 2016). Internalisation includes methods such as "learning-by-doing", simulations, or experiments (Hoffmann, 2009).

Another relevant aspect of the work in OM and IS is the attention devoted to the role played in knowledge management when ordinary technologies are used for daily communication, such as e-mail and video conferencing systems (Bolisani & Scarso, 1999; Goh, 2002). While some studies might be regarded as overly optimistic (e.g. Roberts' (2000) critique of Bolisani and Scarso's (1999) work), much of this research draws attention to the limitations of computer technology for achieving effective knowledge management in companies, even where it is acknowledged to play a vital role (Watson & Hewett, 2006). Thus, Roberts (2000) critically reviewed the potential of then-available ICTs for KES, especially e-mail, voicemail, teleconferencing, video conferencing, desktop-based systems, computer-aided design (CAD), computer-aided manufacturing (CAM), information databases and groupware. Roberts' (2000) assessment was based on findings from studies of knowledge management and transfer in research and development activities in multinational enterprises. The principal discussion focuses once again upon a perceived dichotomy between 'tacit' and explicit or codified knowledge. The author concludes that even if ICTs might facilitate and speed up the collection, storage, and retrieval of knowledge in/from so-called knowledge bases, they would still struggle to convey 'tacit' knowledge. In particular, it is argued that such systems could not translate know-how into show-how (Roberts, 2000).

Studies of coding communities and their development from mailing lists to web-based Q&A help pages stand apart from this discussion. Here, it has been found that Q&A websites are an effective means of knowledge exchange. Based on questions and answers, knowledge exchange is initiated and incentives are created through gamification approaches (Treude et al., 2011; Vasilescu et al., 2013). A positive aspect is that the knowledge-transfer process accelerates development by broadening the knowledge of knowledge seekers. However, participation in the knowledge community represents an interruption of actual coding activities, which can lead to delays. Studies of Stack Overflow have shown that gamification approaches used on the platform work to encourage users to participate and be active in the community, increasing programming productivity (Vasilescu et al., 2014).

Studies in medical settings deal with comparable aspects of expertise sharing in a completely different domain. Normark and Randall (2005) report on the pooling of expertise in an emergency call centre, involving the organised work of several people often at different locations or mobile, who seek and offer relevant knowledge in an appropriate form to meet time-critical conditions. Here, knowledge exchange systems are useful if they support practices that can adapt to organisational needs and be accompanied by two key organisational and cultural changes: (1) developing a culture in which knowledge is organically built and embedded; and (2) developing a strategic role for experts, e.g. shift leaders, who impart knowledge directly or indirectly. Bossen et al. (2013) analysed an IT system that supported home care workers and relatives in cooperative home care work through information exchange and task coordination. Home care is based on a complex network of activities with a characteristically reciprocal division of labour between care workers and relatives. The IT system helped to provide participants with 'peace of mind' that allowed them to realise their individual engagement. The authors emphasised the efficiency of the IT system and noted its strong acceptance among all participants.

In the context of Web 2.0, Morris et al. (2010) explored using social network status messages to ask questions. Their results show that Q&A sites might be more attractive to users with subjective questions on highly personal topics since they afford greater anonymity. By contrast, search engines provide more objective answers to various personal and impersonal concerns. People tend to trust the opinions of people they know rather than strangers, motivating them to turn to social networks. Here, there is a need for rapid feedback and a broad spectrum of high-quality answers. In relation to this, it has been suggested that personalised search algorithms or applications might increase user-friendliness and confidence in the system. The use of social media also creates the possibility of company-wide knowledge exchange in work settings by moving away from intermittent, centralised knowledge management processes to continuous online knowledge-related conversations. How individuals reconcile their work with knowledge transfer during limited working hours is little known, suggesting a need to rethink traditional knowledge-transfer theories and consider other approaches to complex knowledge work. Research here has explored promoting integrating technologies with individual options for action that bring users and technological artefacts closer together using a more "socially acceptable approach" (Majchrzak et al., 2013).

Yang et al.'s (2008) analysis of a Chinese witkey system that connected users via a knowledge-sharing platform and allows a monetary reward for answers holds relevance here. The aim was to explore the underlying properties of users and jobs/tasks to gain insights into participation structures. Their results revealed that monetary rewards were not a significant incentive. By contrast, people's perceived expertise significantly affected participation (ibid.). In another Q&A

related study, Adamic et al. (2008) found clusters according to the length and overlap of requests and replies. While many users broadly contribute across various themes, technical users tend to answer more selectively. Thus, the number of answers marked as helpful is often an indicator of expertise.

2.1.3 The Social Nature of Knowledge and its Relevance in Industrial Contexts

Knowledge is a socially constructed and often distributed asset whose management is deemed an important strategic resource for companies and a source of competitive advantage in the market (Argote et al., 2000; Grant, 1996; Grant & Baden-Fuller, 2004; Levin, 2000; Shingo, 1985; Spender, 1996b; Watson & Hewett, 2006). In CSCW, the focus on issues of knowledge has been placed rather more strongly on the sharing than the management of such an asset. CSCW researchers have long investigated these issues under manifold designations, e.g., OM, collective memory, collective intelligence, expertise location, etc. (Ackerman et al., 2013; Randall et al., 1996) at least since the 1990s. All of these studies examine the role of information in organisational settings, concentrating on the social context and emphasising communication among knowledgeable workers in contexts such as safety-critical and office environments (Ackerman et al., 2013).

The HCI literature has also touched upon the matter, particularly regarding the design of digital visualisation technologies for KES. For instance, Burkhard and Meier (2005) address issues of sharing by introducing two theoretical concepts for knowledge visualisation, based on visual metaphors and knowledge maps. The authors discuss how visual metaphors are powerful for KES but underused in organisations. Their results demonstrate the usefulness of visual metaphors in motivating people to engage in knowledge exchanges and supporting learning processes. In particular, the authors found that visual metaphors support presenting new perspectives, increase remembering, enhance the viewer's focus and concentration, and structure and coordinate communication. While many other studies followed Burkhard and Meier's (2005), most of them have focused on teaching and learning (e.g., (Patterson et al., 2014) (Keller & Tergan, 2005)) or issues of information overload (e.g., (Mengis & Eppler, 2012)) rather than KES within organisational contexts. In manufacturing contexts, even less material has been produced.

The assumption that the primary issue in relation to KES lies in making the previously tacit somehow more explicit has been challenged within the HCI and CSCW (Ackerman et al., 2013; Schmidt, 2012). Research in the field has shifted the emphasis to the social, processual, and other contextual factors that might influence KES. This can effectively thought of as a specific element of the 'turn to practice' characterising much of CSCW work. This re-emphasis is particularly important when the potentially transferable skills under examination are 'embodied', relying on a 'feel' for things (Clarke et al., 2003).

2.1.4 Hardware-Based Expertise Sharing

Several standard procedures exist for supporting expertise sharing within manufacturing settings, including using plans and shift protocols (Carstensen et al., 1999) or coordinating meetings (Schmidt, 1991). However, when focusing on the practice of manufacturing itself, these mechanisms usually fall short because many contextual circumstances are neglected in respect of actual machine operation (Lewkowicz & Liron, 2019). These contextual factors—such as the detailed nature of the materials, the actual condition of the machine (Bowers et al., 1995) or even the current room temperature—can play a decisive role in manufacturing.

Ludwig et al. (2017) primarily focused on these contextual and environmental factors to design a hardware-based approach for collaboratively dealing with this complexity in practice. Using Turkles' (2002) original concept of sociable technologies, they show how Internet of Things (IoT) technologies can offer operators new possibilities for communicating, documenting, and sharing hardware usage practices. Sociable technologies offer three dimensions for understanding a machine's functionality and sharing its practical uses: (1) the internal dimension, which provides the user with information on the inner workings of a machine, its current status, and the functioning of the complex component structure; (2) the spatial-material dimension, which provides detailed information about the machine's location and environment (e.g. room temperature and brightness), which enables deriving possible relationships with the internal context; and (3) the task- and process-related context, which provides information on the position in a production chain or purpose and objective of machine use (Ludwig et al., 2014).

When interacting with specialised machinery, the operator is the central unit (Cimini et al., 2020). With the advent of human-centred manufacturing, new technologies are usually introduced in manufacturing settings to support users rather than replacing them. Although physical demands are changing due to the use of new technologies, assistance systems potentially increase cognitive abilities (Romero et al., 2016). This interaction between the operator, technology, and organisation (Dregger et al., 2016) leads to an augmented knowledge base in which human abilities and technical possibilities complement each other (Engelbart, 1988). Specifically, the ability to interpret data and derive decisions through automated collected data paves the way for extensive expertise sharing (Romero et al., 2016).

2.2 IT Support for Knowledge and Expertise Sharing

The first generation of technical support for expertise sharing comprises repository models, aiming to store and provide all documents in a knowledge domain. The second generation complements the first generation by supporting the communication of knowledge carriers through externalising IT artefacts. An extension of this is the localisation of experts, which is also increasingly supported by IT systems (Ackerman et al., 2013). This paper argues that new technological possibilities are enabling a further generation of technical support for KES, focusing on KES and embodied action as the field of application of technical support is practical work on the shop floor of companies. AI technologies should also be mentioned when looking at the latest developments in IT support for KES. This technology is said to support supervised approaches to KM in particular, as large amounts of data contain patterns that ultimately allow making predictions. AI can therefore address and support different areas of KM, although the focus of AI in knowledge application lies in the area of conversational agents and chatbots (Jarrahi et al., 2023). Embodied AI is an extension of AI technologies in the sense that the data sources do not include images, videos, and texts, but rather data from an environment. This data is processed in simulation environments to replicate the real world. This technology is therefore located entirely in a virtual world and has no direct interface to the embodied tasks on the shop floor (Duan et al., 2022). Accordingly current AI technologies offer few opportunities to support expertise sharing with large proportions of embodied action. This work therefore focuses on technologies at the interface between real data and its virtual processing in CPS in conjunction with a visualisation using AR that is fundamentally suitable for the shop floor. The following chapters present the foundations of CPS and AR.

2.2.1 Cyber-Physical Systems

Existing research has shown that new technologies can potentially support and facilitate numerous different processes, when designed with a view to the social-technical context in which they will be deployed. The rise of AR and sensor technologies provides new opportunities relating to the haptic and thus a more nuanced understanding of the sequencing of activities. When allied with procedures for examining 'rationale', this offers potential for a richer picture of how knowledge and expertise are constituted and thus shareable. Cyber-physical systems (CPS)—as they have come to be termed—have appreciable promise.

CPS are systems of tightly coupled physical and cyber components that integrate software, hardware, sensors, and actuators. The general definition of CPS implies that they can link and manage distributed systems comprises physical and virtual entities (Lee et al., 2015). Interaction with such systems occurs via a human-machine interface, which can be implemented through conventional PC interfaces, touch screens, or even AR-based technologies.

Previous studies have indicated that advances in the area of CPS—especially regarding the integration of AR for visualising particular elements—will bring new opportunities for sharing this kind of embodied knowledge. This work argues that using CPS can mark the beginning of a third generation of KES by overstepping existing assumptions about how knowledge might be gathered and visualised. Thus, CPS might constitute an opportunity to re-vitalise the design of innovative technologies for KES.

In an industrial context, CPS have been termed cyber-physical production systems (CPPS), which have been proposed as a solution to the progressive mechanisation and networking of machines and sensors, resulting in a flood of data (big data) that can be used for various purposes. Under the generic term Industry 4.0, this is seen to represent opportunities for entire industries to position themselves competitively for the future (Kagermann et al., 2013). Among other things, these systems have been considered for predictive maintenance, automatic detection, and remote diagnostic systems (Chen & Tsai, 2017), the development and operational phases in production (Herterich et al., 2015), and the location of trends and the scope for optimisation (Shi et al., 2011). Other research relates to software development for smart factories (Otto et al., 2014) and CPPS-based modelling (Thramboulidis & Christoulakis, 2016).

CPPS comprise autonomous and cooperative elements and subsystems, e.g. smart machines and smart factories, which are connected from the process level to the production level, depending on the situation or context (Monostori,

2014). Their characteristic features include the networking of various production components, such as machines or tools, and the data sets characterising them (Monostori, 2014). Concrete application examples of CPPS exist in both the industrial context and—inter alia—in aerospace, healthcare, civil infrastructure, logistics, military or defence, automotive, energy network, and agriculture (Baheti & Gill, 2011; Rajkumar, 2012; Schuh et al., 2015). A useful overview of their potential and associated challenges can be found in Park et al. (2012) and (Poovendran et al., 2012).

Lee et al. (2015) introduce a structure and architecture for CPS and consequently CPPS, starting with data collection, moving through analysis, up until final value creation. This architecture should serve as a guideline for implementing such systems in industry. These guidelines are supplemented by practical applications and techniques that underpin the theoretical architecture with practical implementation possibilities. Nevertheless, the design and development of useful and usable CPPS has proven challenging, calling upon further research and development in the area.

The broader potential of digital technologies for KES has been widely discussed and exploited within both CSCW and OM (Ackerman et al., 2013; Bolisani & Scarso, 1999; Nonaka & Von Krogh, 2009; Roberts, 2000). However, in the case of CPPS, the burgeoning use of sensor technology offers new opportunities for capturing information about the sequential and haptic characteristics of activities, i.e. the ways in which they are embodied.

The possible application of sensor technology as a central feature of CPS is extensive. For example, in an industrial context, it has the potential to enhance human sensory access by controlling areas that are inaccessible to humans (Chong et al., 2003), and provide additional information (Gellersen, 2005) of assorted kinds. Chen and Tsai (2017) explore using sensors in manufacturing and representative cases of applying RFID and auto ID. The exact nature of deployment substantially depends on the dynamics of the physical process and the degree to which it can be automated. At one extreme, there is a notion of lean infrastructure without external interfaces, where sensors communicate with each other using a knowledge base that contains artefact-based knowledge about the current state of the environment (Strohbach et al., 2004). This requires no human interface because the knowledge base exclusively comprises data and rules for its processing (op. cit.). At the other extreme, there is the notion of a CPS being entirely supervised by humans. Here, a key concern is how data can provide the correct information for the appropriate purpose at the suitable time to the relevant person (Lee et al., 2013).

2.2.2 Augmented Reality

The beginnings of AR technologies date back to the early 1990s (Caudell & Mizell, 1992; Feiner et al., 1993). Their potential for blending useful information with real-world surroundings was already visible from their earliest developments. With the emergence of new technologies such as tablets and smartphones, the popularisation of such applications became a real possibility.

AR has a wide application field, particularly in the industrial context, where there is a huge interest in using AR to provide assistance for complex tasks such as manual assembly or maintenance work (Wang et al., 2016). In these applications, virtual computer-generated visual features are superimposed on the real environment to support executing these complex tasks. These AR applications commonly support workers in sequential tasks by providing context- and task-related information in-situ. Over the last two decades, several studies have investigated the suitability and effectiveness of AR applications for manual assembly or maintenance tasks. In these studies, AR applications were usually compared to paper-based instructions by measuring the time required to complete a task, the number of errors during the task, and the mental workload during the execution.

Classic manuals and standard operating instructions are widely used in industrial activities because they can be standardised. However, these instruments can harm actual work processes as they are often cumbersome and make locating relevant information quite challenging (Norman, 1998). Existing research has demonstrated the potential of AR technology in manufacturing contexts, especially regarding guiding workers in accomplishing tasks such as assembling 3D-structure parts or following the flow of material through manufacturing facilities (Baird & Barfield, 1999; Bhattacharya & Winer, 2019; Dey et al., 2018).

Baird and Barfield (1999) found that AR-based instructions would be more effective for supporting operators with manual assembly tasks. The authors ran a comparative test using four types of instructional media: (1) a paper manual, (2) computer-aided, (3) an opaque AR display, and (4) see-through AR displays. The results of their correlation analysis—considering the assembly time and the number of errors made during the testing tasks—suggested that the AR conditions were more effective in guiding the participants through the test. Despite the variables that might have influenced the study's results—for example, instructions were presented on separate sheets in the paper conditions and different screens in the computer-aided conditions—the findings indicated that the faster assembly times could be associated with the AR condition.

Similar results have been found in more recent research. Henderson and Feiner (2011) investigated the benefits of AR technology in guiding mechanics through maintenance and repair tasks. The authors investigated the extent to which AR applications support workers with tasks concerning the installation and removal of fasteners, indicator lights and connecting cables positioned in a cramped interior. Their findings showed that the AR application used in the study allowed workers to locate the parts and perform maintenance faster. Likewise, Fiorentino et al. (2014) found similar results suggesting that AR-based instructions have a significant impact on workers' overall maintenance-task execution time and the number of errors committed. Echtler et al. (2004) showed that a welding gun modified with AR instructions for position determination supported the user in controlling the positions for the welding spot more quickly. The advantage of this application is that the markup and the welding could take place in parallel.

Funk et al. (2016) compared four different AR hardware types (paper, smart glass, smartphone, and in-situ projection) using a standardised LEGO Duplo task (Funk et al., 2015) with sixteen participants (within-subject). In this study, all hardware types used the same pictorial instructions. The authors measured the task completion time, errors, and task load. The results showed no significant difference between in-situ projections and paper instructions regarding the completion time. However, the same pictorial instruction on an HMD and HHD resulted in a significantly slower time to recognise the correct elements compared to the paper. In addition, more errors were produced when using the HMD compared to the hand-held device and in-situ projection. Regarding the cognitive load, the in-situ projection was perceived to be the lowest and the HMD the highest.

Blattgerste et al. (2018) investigated different in-situ instructions for assembly tasks by comparing four visualisation types (3D in-situ, 2D in-situ, 3D wire, and side-by-side) using the Microsoft HoloLens in a within-subject study comprising 24 participants asked to solve a standardised LEGO Duplo task (Funk et al., 2015). While the results show a faster completion time and fewer errors when using 3D in-situ visualisations, there was no significant difference between the systems in terms of the task load. Blattgerste et al. (2017) further compared different types of AR hardware (Microsoft HoloLens, Epson Moverio BT-200, and smartphone) to paper-based instructions. The paper-based instructions resulted in the fastest overall task completion time, while the Microsoft HoloLens led to the lowest number of errors, albeit with a significantly higher cognitive load.

Smith et al. (2020) analysed the effect of different interaction modalities (touch and voice) and visualisation modes (3D model, text annotation, and in-situ video) of an AR application on task load, task completion time, and error rate for 24 participants performing a LEGO Duplo task. The in-situ video led to the fastest task completion time, followed by superimposed 3D models, and text annotations. In addition, the task load was the lowest for the in-situ videos and the highest for the text annotations. The method through which the users interacted with the AR content (touch versus voice) had little to no effect on the measured performance.

Radkowski et al. (2015) investigated the effect of different visual features (abstract versus concrete) of 3D content for various degrees of difficulty in manual assembly tasks. Abstract refers to visual forms (e.g., lines, colours, shapes) separate from any concrete 3D models. The authors conducted a between-subject study with 33 participants (abstract visualisation versus concrete visualisation versus paper-based) who had to assemble a mechanical axial piston engine with sixteen manual assembly process steps. Two of these steps were rated with a high degree of difficulty, while all others were rated with a low degree. Compared to the LEGO Duplo tasks (Funk et al., 2015), this assembly task can be considered more difficult as it is based on real conditions. The authors used a tabletop AR workstation displaying an AR view of the working space. They showed that the abstract visualisation led to a longer completion time and more errors than concrete visualisations. Further, the concrete AR visualisation resulted in a similar average completion time compared to the paper-based instructions, while the completion time for abstract visualisation was almost twice as long. This result suggests that concrete visualisation is more suitable for relatively simple tasks. However, it could not to determined which kind of visualisation type is more appropriate for rather complex tasks as there were ambiguous results regarding the time of the two difficult assembly steps.

Wiedenmaier et al. (2003) argue that users only benefit from AR when focusing on difficult assembly tasks, and most previous studies (Blattgerste et al., 2017; Funk et al., 2016; Radkowski et al., 2015; Smith et al., 2020) have confirmed that AR does not provide an advantage for the completion time of simple tasks. Haller (2004) concluded that 3D models look nice for AR developers, although the actual end users often ask for abstract but familiar schemes or 2D drawings. These statements contradict the results of Blattgerste et al. (2018) and Radkowski et al. (2015).

Many other studies have shown similar outcomes (Dey et al., 2018). One potential reason for improved performance with AR support is that multimodal feedback (e.g., visual plus auditory or tactile feedback) can be more effective than visual feedback alone (Burke et al., 2006). In turn, Tang et al. (2003) provided evidence that AR technology can reduce cognitive load and therefore the possibility of overload. Feedback provided by AR systems partly differs from classical systems intending to adjust the action. Digital technology provides constant, real-time updates on progress (Goomas, 2012; Hermsen et al., 2016), enabling more effective immediate feedback (Goomas et al., 2011). Research has shown that feedback on the state of a machine—for example—can be processed particularly well if the feedback and the positioning of the information move with the machine (Klinker et al., 2004). Evidence has also shown that AR technology can increase accuracy and consequently reduce workers' cognitive load (Tang et al., 2003). several authors have argued that inaccuracies might increase excessive extraneous cognitive load, which can lead to overload within a restricted working memory system (Bresciani & Eppler, 2009; Paas et al., 2003). AR can potentially minimise complexity by adjusting reality to virtual representations and eliminating the need for spatial placement.

Despite the promising results of various studies, some researchers have highlighted AR's limitations. Regarding accuracy, Kosslyn (2006) has drawn attention to the lack of unquestionable results regarding AR-based visualisation's impact on accurately positioning the visualised elements. Studies such as Werrlich et al. (2019) have also checked some claims concerning the potential support of AR for assembly workers. In their study, the authors' results suggested that assembly workers would complete assembly tasks much faster with paper instructions in some situations. Nevertheless, the authors did find that the number of errors committed when using AR support was lower than when using paper-based instructions (Werrlich et al., 2019). Blattgerste et al. (2017) derived similar results, contrasting with the results of Funk et al. (2016), who found evidence that AR-based instructions presented on an HMD could be associated with a higher number of errors and significantly higher cognitive stress. The same applies to Büttner et al.'s (2016) study, showing that AR with an HMD performed worse than projection-based AR and paper due to the lack of technical hardware equipment at that time. The types of hologram visualisation also influence performance, with recent research showing that concrete in-situ holograms positively influence performance (Blattgerste et al., 2018).

In terms of applications, much existing research has focused on mobile AR applications to support synchronous remote or co-located collaboration. These cases commonly anticipate the in-situ creation of content and its distribution among the actors (Henrysson et al., 2005; Langlotz et al., 2010). In keeping with this trend, Gauglitz et al. (2012, 2014a, 2014b) have examined the scope for telecollaboration mediated by AR technology. In their work, sophisticated tracking algorithms have been devised to help remote users to place location-based annotations without marker recognition in the field of view of a local user. Based on this, novice users could follow instructions from experts who were remotely placed. The evaluation showed that the proposed technology could aid novice users in following complex instructions quickly and with almost no errors (Gauglitz et al., 2012). Further developments of this application enabled remote users to act independently of the field of view of the local users, while still placing location-related annotations. Additional results indicated that the proposed solution was both easy to use and provided significant help in fulfilling the tasks in the described context (Gauglitz et al., 2014b).

The use of AR in knowledge-intensive environments to support learning processes has also been explored to some extent. A promising feature is fast, unrestricted, and context-specific access to information (Klopfer et al., 2005). This type of technology allows for the following affordances, which are also likely to hold use in an industrial context: (1) exploring teaching materials from different angles; (2) supporting teaching where no real-world first-hand experience is possible; (3) enhancing collaboration between students and instructors; and (4) helping students to control their speed of learning and decide which pathway to choose (Yuen et al., 2011). Findings also suggest that AR technologies can enhance workers' skills and perception, promising potentially sustainable improvement in tasks such as assembly operations (Ong et al., 2008; Pathomaree & Charoenseang, 2005).

Unsurprisingly, embodied action has arisen as an issue in this area. For instance, Fakourfar et al. (2016) highlight the role of deixis in AR systems, i.e., the use of gestures in the course of interaction, noting that deixis facilitates collaboration by allowing for common reference, thus acknowledging the importance of the bodily and spatial aspects of human action. This gives particular substance to the relevance of embodied action to KES. Fakourfar et al. (2016) stress that AR annotation systems must accommodate deixis, even if this leads to further challenges. Irlitti et al. (2017) comment on how the combined shared space provided by many AR systems allows people to use the same non-verbal cues that they use in face-to-face interaction. Tang et al. (2007) explore the role of embodiment in particular detail in the use of mixed presence groupware, i.e., groupware supporting interaction between groups where some people are co-located and others are remote. The authors note that when such systems do not account for the embodied actions during collaboration, it leads to a phenomenon known as presence disparity. The authors therefore suggest the need to provide mechanisms for embodiment in mixed presence groupware that can offset presence disparity.

Results have shown that an AR-supported learning process has no significant advantages over conventional instructions in terms of the number of errors, execution time, and workload. Results from previous studies have suggested that the learning curve flattens after the initial training phase (Wilschut et al., 2019).

In a study on training analysis, Büttner et al. (2020) also showed that personal training (one-on-one training with an expert), training based on an in-situ projection, and training based on a paper instruction performed equally well in the short term, with neither training cycles nor training times significantly differing. The second training day showed different results, with personal training outperforming the other training modes. The application of the learned content after 24 hours showed an identical picture. Here, again personal training was significantly better, although this effect was no longer noticeable after one week. Using AR in a learning environment to explain complex machines has proven worthwhile as the machines can be fully explained via the new interface and the handling can be instructed. An interface for data transfer between AR visualisation technology and the machine enables the user to see the machine's values directly in the AR interface and thus better understand the machine (Eder et al., 2020; Jasche & Ludwig, 2020).

Projection-based support systems are increasingly used in the industrial environment. These systems can improve classic instructions by guiding users without prior knowledge through the assembly task (Rodriguez et al., 2015). In addition to the findings that these systems perform better than HMD (Büttner et al., 2016), recent research has shown that the appropriation of operating complex machines is supported by projected in-situ videos, thus blurring the boundary between explicit and implicit knowledge. A further advantage of projection mapping is that it supports appropriation accompanied by other people (Ludwig et al., 2019). In addition, the projection mapping's necessary sensory equipment allows a direct check of the assembly results and thus promotes the users' well-being (Ojer et al., 2020). However, more critical results from studies with in-situ projections have also been reported, indicating that not all study participants improved their performance with this support, whereby the exact causes remained unclear (Korn et al., 2013).

2.2.3 Augmented Reality Authoring

One demand in manufacturing contexts is to keep documentation of work processes and practices up to date (Henderson & Feiner, 2011; Radkowski et al., 2015). However, existing has found that such documentation is usually outdated. One possible reason is that work practices and processes evolve with the time and availability of new resources, and the production of new documentation does not keep pace (Novick & Ward, 2006; Randall et al., 1998).

The literature has found that producing AR-based content can be challenging, expensive, and time-consuming (Bhattacharya & Winer, 2019). Neumann and Majoros (1998) argued for elaborating an AR-media language to facilitate the creation and deployment of AR content across platforms. According to the authors, content should be built as media that could be visualised in different contexts. This must represent geometric data in the form of CAD models (Neumann & Majoros, 1998). More recent research has shown that holograms are the best fit for KES in an environment of complex tasks (Kosslyn, 2006; Speier et al., 2003). They can account for the position of information to be presented and the displayed object orientation (Newell et al., 2014; Rzayev et al., 2018).

Existing authoring tools focus on transferring first-design prototypes from paper or plasticine into AR content (MacIntyre et al., 2005; Nebeling et al., 2018; Nebeling & Madier, 2019; Nebeling & Speicher, 2018). Thus, new applications have aimed to support the generation of AR content from simple 2D sketches and plastically modelled objects without having relevant programming knowledge (Nebeling et al., 2018). Nevertheless, challenges still arise with enabling end users to create and share AR content in their normal work environment without repeated interruptions, which might affect task performance (Nebeling & Speicher, 2018; Speier et al., 2003). Hence, the increasing attention focused on new approaches to create technical documentation is unsurprising (Bhattacharya & Winer, 2019).

The most conventional approach arguably comprises collecting summaries of existing instructions in a knowledge base over a long period. These summaries are posteriorly edited using a GUI editor and displayed in an AR view using object-recognition algorithms (Jo et al., 2014). Another traditional approach is manually creating instructions via 2D authoring wizards, which can be posteriorly displayed in a mixed-reality environment (Zauner et al., 2003). More recent approaches have tackled creating complex AR instructions based on existing 2D documentation using 3D models (Mohr et al., 2015). The proximity to 3D modelling tools is deliberately sought to make the authoring environment as intuitive as possible (Seichter et al., 2008). In addition, sensory values and complex recognition algorithms can be used to derive AR instructions when performing a task. Subsequently, they must be manually refined before novices can use the instructions (Agresti & Finlay, 2008).

The literature has shown that intelligent authoring environments have advantages in executing assembly tasks. In addition to the pure generation of virtual content, ontologies and logic are stored in the authoring environment, enabling intelligent guidance and feedback on the performed task (Westerfield et al., 2015). Authoring environments implemented via a desktop application are preferred in specific stationary scenarios as users perceive them as more performant (Gonzalez et al., 2019). In such scenarios, tangible AR authoring interfaces are also listed as an intuitive environment for creating AR content (Lee et al., 2004).

Using sensory information and automating parts of the AR content creation aligns with a CPPS-based approach to KES. CPPS hold strong potential to ensure the networking between production plants with product design, supply chain management (SCM), and quality assurance processes through the provision of interfaces to the production plant and the collection of machine and sensor data (Wagner, 2018). In terms of KES, researchers have argued that AR-based CPPS are about to bring significant changes in how knowledge and expertise are documented and shared between knowledgeable actors through support sharing specific forms of embodied knowledge. While sensors and other types of technologies and computations can potentially record information about specific actions often overlooked, AR technology can mix this information with the real world to convey a richer representation of specific processes (Leit et al., 2016; Pacaux-Lemoine et al., 2018; Wang et al., 2014). In particular, they can offer new ways to upskill workers by facilitating creating instructional material, which can be recurrently used in various situations. Additionally, sensory equipment immediately detects failures, allowing timely intervention (Funk et al., 2017). Ludwig et al. (2016) showed that the sensor captures within CPPS are also suitable for capturing contexts and environmental variables, thus providing a better understanding of production processes.

2.3 Research Gap

After reviewing the relevant findings from the research field of CSCW and especially regarding KES, introducing essential technical concepts such as CPS and AR, and providing a basic classification of set-up processes, this chapter identifies the research gaps. For this purpose, this chapter derives the overarching research goals by showing that new forms of KES in the industrial context and particularly in industrial set-up processes are enabled based on modern technologies, whereby these possibilities have not yet found adequate grounding in science and literature.

The application field of industrial set-up processes has been intensively optimised and scientifically illuminated by various methods borrowed from lean management. The findings focus on the single minute exchange of die (SMED) method (Dillon & Shingo, 1985) and its successful application in different contexts (Cakmakci, 2009; Ferradás & Salonitis, 2013; Palanisamy & Siddiqui,

2007). Van Goubergen and Lockhart (2005) highlight that the critical role of people and their level of knowledge remains largely unexamined. Communication and decision support systems and their technical implementation are largely unexplored in the context of set-up processes. Therefore, set-up processes must be fundamentally rethought by characterizing them from the human perspective under the aspects of knowledge and expertise sharing. This leads directly to the first research question. *How can set-up processes be characterised?* A key finding from the empirical research in this thesis is that set-up processes can be divided into static and dynamic processes (see also Chap. 6). The following research questions directly address this new characterisation into static and dynamic set-up processes.

In the tradition of CSCW, many studies have been conducted in the environment of production areas (Bowers et al., 1995; Carstensen et al., 1999; Clarke et al., 2003). Given the need for KES to evolve in production areas (Ackerman et al., 2013), the accurate capture, description, and analysis of the KES process in the context of production-related activities is another milestone in the further development of KES. Existing work in the field of KES represents its different generations, and significant progress has been made in understanding knowledge and competence exchange. However, changing technologies prompt new opportunities to support KES. At present, it remains unclear how these technologies contribute to successful knowledge transfer on the shop floor, especially as it is important to transmit non-propositional knowledge through observing embodied actions in an industrial context. The tacit knowledge from the OM and IS literature led to a debate in the CSCW literature about the specifics of people's practical skills and knowledge in performing certain tasks (Ackerman et al., 2013; Schmidt, 2012). This debate is relevant to the interest in production processes because it addresses how production activities should be performed, represented, and shared with others who might want to engage in similar activities like set-up processes. The central challenges and implications of set-up processes therefore need to be investigated more closely from the perspective of KES. The second research question addresses this aspect.

CPS are particularly interwoven with cooperative activities and computation. However, so-called 'human-in-the-loop' CPS bring immense challenges to the design of such systems (Schirner et al., 2013). The interplay and intertwining between physical and virtual environments is particularly challenging, requiring novel tools for the conceptualisation and user-centred and participatory design of such technologies (Lee, 2008; Monostori, 2014; Paelke & Röcker, 2015). To date, it has also been a major challenge to make CPPS and Industry 4.0 concepts accessible to small and medium-sized enterprises (SMEs; Ludwig et al., 2016).

Considerations of explicitly building knowledge-sharing support into CPPS are largely absent from the literature. Lee (2008), Monostori (2014) and Paelke and Röcker (2015) have discussed some of the challenges associated with the design and implementation of CPPS, usually described in terms of increased technical complexity, the need for new interaction concepts that can consider the unpredictability of interaction between physical and the virtual worlds, and the lack of prototyping and test tools. While all of these challenges are relevant to research, they remain underexplored (Schuh et al., 2014). In particular, qualitative insights exploring contextual variation and the need for better tailoring of solutions are difficult to find. Together with the potential advantages of CPPS—e.g., enhanced productivity and flexible production systems (Monostori, 2014; Paelke & Röcker, 2015; Post et al., 2017)—this trend underscores the timeliness and relevance of this contribution. In view of CPPS's potential for recording the haptics and detailed variables involved in work processes that can be visualised through AR-based technologies, it is conceivable that such systems might prompt a methodological revolution in the gathering and disposition of knowledge, including the knowledge embedded in embodied action.

In CPPS, AR technology is simply part of a broader construct that integrates sensor technology and extends beyond instructions to span manufacturing processes' logistics and the quality control of resulting products. However, AR technology plays an essential role because it can help to better integrate data into the real context (Leit et al., 2016; Pacaux-Lemoine et al., 2018; Wang et al., 2014) and thus has the potential to support knowledge-intensive processes. Despite Tang et al.'s (2007) work, there is no discussion in the literature regarding how embodied action captured through AR systems can be used for KES. Furthermore, little attention has been paid to scenarios of asynchronous collaboration (Irlitti et al., 2017), which is core to this research. This work addresses this literature gap and contributes towards reducing it. In addition, essential questions regarding an authoring environment adapted to the industrial environment for creating AR content remain open (Nebeling & Speicher, 2018). In this context, the combination of AR technology and an extended sensory recording of the environment plays an important role in enabling recording non-obvious environmental influences (Ludwig et al., 2017) and allowing the AR technology to place this information in the real environment (Leit et al., 2016; Pacaux-Lemoine et al., 2018). The integration of both technologies and their embedding in a KES process on the shop floor has not previously been highlighted. The requirements for support must therefore be researched using new AR-based CPPS and the resulting possible support mechanisms for people must be evaluated. The third research question addresses the requirements and the fourth and fifth research questions address the possibilities of support for both static and dynamic set-up processes.

In summary, the following five research questions can be derived:

- How can set-up processes be characterised?
- What are the central challenges and implications for the technical support of knowledge-intensive static and dynamic set-up processes?
- What are the requirements for an implementation to support static and dynamic set-up processes?
- How can static set-up processes be supported by AR-based CPPS, and how should holograms be designed to support them?
- How can dynamic set-up processes be supported with the help of AR-based CPPS?

3 Research Approach

3.1 Methodological Approach

To understand and analyse the challenges and details of the set-up processes, various empirical studies were conducted in different companies in two European countries. The studies always focused on the activities of the setters and machine operators on the shop floor. Formally, set-up processes start after the production of one article and finish before the production of a second article starts (Van Goubergen & Van Landeghem, 2002). However, Van Goubergen and Lockhart (2005) highlight that the role of the person has a significant influence besides all formal specifications regarding a set-up process. Therefore, the set-up processes should not only be considered from a technical and organisational perspective but rather the social practice of set-up should be the focus of the work.

For this reason, the design case study (DCS) provides a central framework as IT artefacts are designed, implemented, and evaluated based on the understanding of social practices (Wulf et al., 2015). The approach is characterised by three main phases. In the first phase, the users' practices are brought into focus with the goal of gaining an understanding of them (Wulf et al., 2015). For this purpose, qualitative empirical methods are used in the context of the thesis. Workplace studies based on shadowing sessions hold central importance (Czarniawska, 2007). In these sessions, in-situ data of workplace practices and the accompanying social interactions were collected and extended by eye-tracking sessions. For this purpose, eye-tracking glasses were placed on different set-up workers and set-up processes were recorded. The operators were instructed to use think aloud (Nielsen, 1993) to record not only the execution of the action but also the motives and background knowledge. Based on the knowledge base

S. Hoffmann, *Knowledge Sharing on the Shopfloor*,
https://doi.org/10.1007/978-3-658-50133-4_3

gained from this, semi-structured interviews were conducted with different people involved in the set-up process. These interviews and the audio tracks of the eye-tracking recordings were transcribed using the intelligent verbatim method (Hickley, 2003). The eye-tracking data was analysed in further depth using Tobii software and heat maps were created. The transcripts generated were subjected to thematic analysis (Braun & Clarke, 2012).

The second phase of the DCS deals with designing the new solution based on the requirements identified in the first phase. The procedure should be characterised by a participatory approach (Wulf et al., 2015). Specifically, design workshops were held with operators and other people involved in the set-up process. The results of the workshops were consolidated in prototyping activities and summarised in design concepts. The design workshops follow a scenario-based approach to design (Carroll, 2000). In addition to sketches and whiteboard recordings, audio recordings of these workshops were also conducted, before being transcribed and subjected to thematic analysis (Braun & Clarke, 2012). Wulf et al. (2015) highlight that it is not a matter of strictly sequential phases and activities within the phases but rather an iterative procedure and a repeated running through of the phases with the insights already gained. This is reflected in the design phase and activities carried out, which not only included design workshops but also medium- and high-fidelity prototyping with subsequent initial evaluations. Heuristic evaluation (Nielsen & Molich, 1990), cooperative evaluation, and lab usability tests were conducted.

The final evaluation phase includes research into the introduction of the IT artefact, its appropriation in the practice environment, and the subsequent artefact redesign. The iterative approach between the individual phases of the DCS holds central importance here, in the sense that the findings from the third phase find their way back into the second phase (Wulf et al., 2015). The evaluation of the set-up support took place in different companies on very different machines. Setters and machine operators had the task of creating set-up processes and carrying them out based on the instructions presented. During the evaluations, observations were made and documented with notes, as well as interviews conducted with the persons involved. Both the notes and the interview transcripts were subjected to thematic analysis (Braun & Clarke, 2012). In addition, quantitative data was collected through a laboratory study to better address performance aspects of AR instructions.

3.2 Research Field

The research field covers set-up processes as an essential and success-critical part of production processes in manufacturing companies. Set-up processes include preparatory activities on a machine or tool before the actual production can start, which are usually carried out after the production of one article and before the start of production of a second article (Van Goubergen & Van Landeghem, 2002). In most cases, the start of production directly depends on a successfully completed set-up process, making it a critical process in production. Moreover, set-up times during which the machine is at a standstill are not productive time. Therefore, the goal must be to design set-up processes as efficiently as possible in terms of economic production. Efficient implementation is particularly desirable, especially against the background of small batch sizes, short delivery times, and the associated increased number of set-up processes (Van Goubergen & Van Landeghem, 2002).

Since the economic efficiency of set-up processes is playing an increasingly important role in production processes, Dillon and Shingo (Dillon & Shingo, 1985) developed a method called SMED, which aims to optimise set-up times and therefore minimise the time between the last good part of the old job and the first good part of the following job. An essential instrument is the so-called internal and external set-up. Each individual step of a set-up process is evaluated and a decision is made concerning whether the step must be carried out with the machine at a standstill (internal set-up) or whether the step can be carried out with the machine running (external set-up). In principle, the aim is to convert all internal set-up steps into external set-up steps (Dillon & Shingo, 1985). In addition, further technical and organisational measures can shorten set-up processes (Conrad, 2016). Van Goubergen and Van Landeghem (2002) describe that the quality of a set-up process is influenced by technical equipment and tool aspects, the organisation of the work, and the method used. Cakmakci (2009) takes up the latter point and explains that SMED approaches and experiences already provide support by designing the tools in a set-up optimised way, thus helping to reduce the set-up times in the operative set-up process. Furthermore, Van Goubergen and Van Landeghem (2002) take up the SMED approach and propose three steps to reduce set-up times. First, all set-up steps are evaluated to ascertain whether the machine needs to be stopped to perform the step. If steps can only be carried out when the machine is at a standstill, an attempt should be made to change these steps so that they can also be carried out when the machine is running. Finally, the sequence of the steps should be optimised so that an optimised set-up sequence results. Overall, the set-up instructions play a key role since the authors consider them to be an essential specification for executing the set-up steps. The aim is to provide the operators with standardised and optimised set-up instructions as this increases the probability of short set-up cycles from the outset. In the literature, there are various examples of the application of the SMED system and technical extensions of the SMED system. It emerges that the SMED system leads to improvements in the set-up processes (Cakmakci, 2009; Ferradás & Salonitis, 2013; Palanisamy & Siddiqui, 2007). However, there are also critical views claiming that important other measures are not recognised or even measures are carried out in the incorrect order due to formalising the process (McIntosh et al., 2001).

The human factors of influence remain largely unconsidered in all systematic approaches related to SMED. Van Goubergen and Lockhart (2005) highlight that people play a critical role in the set-up process, whereby their knowledge has a significant influence on the processes. However, significant human factors of influence remain largely unconsidered, including communication aids and their design for exchange during set-up. In addition, tools to support decisions are not sufficiently considered. This is especially the case when existing instructions need to be optimised based on experience. Furthermore, the presentation of instructions and the design of displays are not sufficiently considered. Therefore, these factors must be included in the set-up optimisation process, and design recommendations must be elaborated (Goubergen & Lockhart, 2005).

Detailed research into knowledge management practices in set-up processes must be carried out on the shop floor of manufacturing companies. As part of the subsequently listed research and transfer projects, set-up processes were therefore empirically analysed at various companies. A variety of production processes and machines were used in the companies, although the main focus was placed on tool-based metal forming processes. Collaborations between the companies took place, and a large number of people were involved in the set-up processes. In addition to operators, setters, foremen, master craftsmen, sales staff, construction engineers, quality engineers, and production engineers were also involved. Furthermore, the people involved had different job tenures, which indicates their respective experience. The companies and the people involved are listed separately in the individual publications.

The results of the work essentially stem from two research projects. First, from the research project entitled 'Cyber Setup 4.0: Cyberphysical support of humans during the set-up process using the example of a bending process for small batch production based on a knowledge transfer approach', funded by a grant of the European Union and EFRE.NRW (No. EFRE-0800263). During the research project, set-up processes were empirically investigated from 2016 to

2019, and support options were implemented and evaluated based on modern AR technologies. The focus of the project was placed on supporting static set-up processes in the form of assembly and disassembly work and their instruction with the help of the aforementioned technologies.

Second, implementation projects were carried out as part of the Mittelstand 4.0 Siegen Competence Centre, part of Mittelstand-Digital. With Mittelstand-Digital, the Federal Ministry for Economic Affairs and Climate Action supports digitisation in SMEs and the craft sector. Empirical studies and the resulting implementations and evaluations of AR software were conducted as part of the implementation projects. The focus here was placed on the support of dynamic set-up processes, which include setting parameters as part of the set-up process.

Part II
Findings

4 Of Embodied Action and Sensors: Knowledge and Expertise Sharing in Industrial Set-up (JCSCW 2018)

Knowledge and expertise sharing has long been an important theme in CSCW and, importantly, one that has frequently challenged a prevailing view concerning knowledge management. This critique focused, initially, on the practical problems associated with issues of Organisational Memory (OM), and in particular the difficulties inherent in an oversimplified 'repository' model. Attention then turned to issues of contextuality and communication for expertise sharing, drawing on concepts such as communities of practice and social capital to understand, again, the sharing of knowledge and expertise in practice. Here, we report on how particular kinds of 'embodied action' can be identified in relation to the potential of cyber-physical infrastructures for knowledge sharing in an industrial context. We argue that, in a complex industrial domain, both the recording of physical movement—'showing'—and the representation of local knowledge—'telling'—are potentially relevant. Our proposal is that the evolution of cyber-physical infrastructures now offers a way of changing some early assumptions about how knowledge might be captured and displayed. We argue that we are entering a third generation of knowledge and expertise sharing research, where the use of augmented reality (AR) and sensor technology will result in significant new methodological innovations, including the capture and sharing of knowledge, embedded in embodied action.

4.1 Introduction

Since its inception, a classic problem in CSCW has been how to capture, describe and demonstrate real-world knowledge and skills so that this can be drawn upon by other people. The concern is shared with some other domains as well, notably the field of Knowledge Management (KM), where the focus has been on practices

S. Hoffmann, *Knowledge Sharing on the Shopfloor*,
https://doi.org/10.1007/978-3-658-50133-4_4

associated with managing knowledge work. This includes not just the service sector, with highly skilled professionals working with intangible goods, but also manufacturing, especially advanced manufacturing and the production of complex products, where blue-collar workers are often actively engaged in planning and coordination (Nonaka & Takeuchi, 1995; Schmidt, 2012).

Attempts to understand how knowledge and expertise can be shared have generated an abiding interest in the ontological and epistemological dimensions of knowledge. It was Ryle (Ryle, 1945, 2009) who famously distinguished between 'knowing how' and 'knowing that'. The point of this distinction was to show that:

> *"When a person knows how to do things of a certain sort (e.g., make good jokes, conduct battles or behave at funerals) his knowledge is actualised or exercised in what he does. It is not exercised … in the propounding of propositions" (op. cit., p.8).*

'Knowing that', by contrast, is perfectly open to propositional expression; for instance, knowing that it takes 2 hours and 15 minutes to travel by train from London to Paris can be represented in a timetable. 'Knowing how', on the other hand, requires no theoretical understanding but simply describes how the body acquires certain learned habits. This distinction has occasioned considerable debate (Ackerman et al., 2013; Schmidt, 2012), not least in the KM literature (Hager, 2000; Howells, 1996; Spender, 1996a; Tuomi, 1999).

Our intention in this paper is not to extend these debates. However, it is worth noting that, following on from Polanyi (Polanyi, 1967), the notion of 'tacit knowledge' became fashionable in a number of quarters. In their discussion of a 'knowledge-creating company', Nonaka and Takeuchi (Nonaka & Takeuchi, 1995) attempted to operationalise this concept for KM purposes. Since then it has come to dominate in much of the KM literature. However, such attempts, at least sometimes, appear to mystify the idea of judgement or indeed, practice, through what, in our view (and others—see (Schmidt, 2012)) are spurious appeals to the 'tacit'. The point would be that, if one accepts that certain kinds of knowledge have this mysterious quality, then methodologies for making it explicit and formalisable are needed.

We (as many others) have some serious reservations about this view. Our own studies indicate that much so-called 'tacit' knowledge is not tacit at all and is regularly articulated in conversation or manifested in gesture, glance, gaze and so on. It is, however, done in highly contingent and local circumstances (see e.g. (Randall et al., 1996)). The fact that 'know how' has no propositional content does not mean it cannot be described, understood or learned, or that such descriptions will not sometimes need translation into propositional statements. However, such descriptions will, as and when they are deployed, be contingent and to a purpose. Further, and as we shall see, there is no particular reason to assume that non-propositional knowledge has to be translated into a propositional form. This, of course, raises a further issue, which has to do with the 'local' character of some knowledge. As has been pointed out on a number of occasions, not least by parties to this paper, much work practice remains 'invisible'. 'Know how' may well be important to participants to a work activity, but its importance may not be recognised by others. In contrast to the conventional KM view—that the key issue is how to formalise the tacit through documentation and other constructs or, through more mundane methods—we identify just how (typically) non-propositional knowledge and, moreover, knowledge which has a 'local' character, may be deployed with a view to understanding how it may be displayed using more innovative technologies. More to the point, however, the various kinds of possible support will need to be configured so as to reflect the different kinds of knowledge and expertise being displayed and the different modes in and through which they are displayed.

Hager (2000), has made the point that investigating 'know how' carries with it certain assumptions about the need for qualitative approaches. Our own work, alongside that of many others in CSCW and HCI (see e.g. (Bowers et al., 1995; Clarke et al., 2003; Harper et al., 2000; Luff et al., 2000)) testifies to the fact that local, 'tacit' and otherwise 'invisible' work practices[1] can be rendered visible (largely unproblematically) through careful observation. Nonaka himself (Nonaka, 1991) notes how 'tacit' knowledge concerning bread kneading can be grasped by an apprentice through close observation of the work practices of an expert bread maker.

[1] This is not to conflate 'invisible work' with 'tacit knowledge', though it has been argued in the literature that tacit and contextual knowledge is one of the 'many guises' that invisible work can take ((Nardi & Engeström, 1999), p. 2). To clarify, invisible work has been discussed in the literature as: work that happens behind the scenes; routine or manual work entailing unacknowledged problem solving; work undertaken by people who remain unaccounted for; or informal work that is outside any job description, yet vital for collective tasks and company organization (Nardi & Engeström, 1999). In turn, 'tacit knowledge' has been mooted to be: 'non-codified knowledge' acquired through learning-by-doing (Roberts, 2000); 'highly personal' and 'difficult to formalize' ((Nonaka, 1991), p. 98); 'deeply rooted in action' and 'context' (Matthew & Sternberg, 2009; Nonaka, 1991), 'not yet … abstracted from practice' and 'applied in the state of the flow' (Spender, 1996a). In Argyris' (2021) view, constant use of a particular instance of knowledge leads to internalisation and its automatic use. It thus becomes understood or implied without being articulated, and as such, is usually taken for granted and goes unnoticed (Baumard, 1999).

Nevertheless, there is also agreement that it is by no means trivial for workers to capture for themselves these kinds of practices or the non-propositional knowledge that might be embedded in them. A key issue here is time. Workers are typically under pressure to achieve a range of goals that do not allow much time for reflection. This being the case, the overhead created by knowledge and expertise sharing activities can appear, at least at first, to be overwhelming (Ackerman & Malone, 1990; Schmidt, 2012). A second issue relates to privacy and trust and the desire to retain one's competitive edge and status within a company. Thus, some workers are resistant to the idea of sharing their knowledge—especially knowledge acquired through hands-on experience over many years (Brown & Duguid, 1998; Nonaka, 1994). Beyond this, there is also an issue of having what Schmidt describes as the 'expositional and didactic competencies required to formulate their experiences so that it is understandable for a remote and possibly unknown readership' ((Schmidt, 2012), p. 206). In other words, it is hard to express 'know how' in a generic form such that just anyone might be able to make sense of it.

Despite our apparently critical stance, we acknowledge the advances that have taken place in knowledge and expertise sharing over the past few decades. Ackerman and his colleagues, for instance, successfully addressed a number of key issues for knowledge and expertise sharing through their implementation of Answer Garden (AG) 1 and 2, which built upon tools that had already been designed to support certain aspects of organisational memory (Ackerman & Malone, 1990; Ackerman & McDonald, 1996). Other tools have followed in the wake of these early endeavours (see (Ackerman et al., 2013), for a review of the more important developments). However, the key issues outlined above, we suggest, may be open to solutions that the CSCW community has not yet fully explored.

On occasion, new technologies bring with them significant new affordances. Recent advances in the capacity of AR-based cyber-physical systems (CPS), put together with the potential opportunities offered by the Internet of Things, are about to provide just this kind of a technological step-change. This, we suggest, will bring with it new ways in which knowledge and expertise might be captured and shared. Put another way, AR provides insights into how the cyber-physical innovation might afford the sharing of knowledge and expertise. This relates especially to local and embodied knowledge, which, setting aside the various debates, is generally acknowledged to be difficult to capture and share (Ackerman et al., 2013; Fitzpatrick, 2003).

If we accept the above arguments, it follows, pace Nonaka and Takeuchi, 'know how' does not necessarily need to be transformed into propositional content for effective knowledge sharing to take place. A range of other possibilities might be considered. Amongst these is the ostensive provision of information about 'know how' by using video, virtual reality and—increasingly, as the relevant technologies mature—AR and sensor technology.

In this paper, we elaborate on some initial ideas about the use of CPS and AR technology that have already been discussed in CSCW (Ludwig et al., 2017; Robertson & Wagner, 2015), with the specific goal of advancing the state of the art concerning knowledge and expertise sharing. We do so by reporting on ethnographic data collected in an industrial context (itself under-reported in the CSCW and HCI literature), with the aim of identifying some of the embodied 'know how' that could fruitfully be shared for one specific activity: set-up for bending machines[2]. This has been selected because of its perspicuous character for investigating the potential role of sensor technology in the representation of workplace practices, as will be later clarified. To the best of our knowledge, articulation of the use of CPS for knowledge sharing in industrial set-up has not been addressed in the CSCW or HCI literature. These three factors, together with our discussion of how this instantiates a possible new approach to knowledge and expertise sharing, are the key contributions of this paper.

The paper is structured as follows: First of all, we examine related work, looking in particular at how knowledge sharing has been discussed in various domains. After this we present our research in more detail and our methodological approach. We then look at the empirical results, focusing on the relevance of knowledge to industrial set-up processes, and the difficulties entailed in sharing it. After this we introduce some design implications that need to be born mind when seeking to support knowledge and expertise sharing in these kinds of industrial contexts. This is followed by a discussion of the findings and design implications, elaborating how CPS-related AR and sensor-based technologies might fuel a step-change in knowledge and expertise sharing support. Finally, we conclude by pointing to new directions that might fruitfully be pursued in knowledge and expertise sharing research.

4.2 Related Work

Aiming at setting the background for the later discussions in the paper, we start this section by revisiting some of the key texts concerning the knowledge-intensive character of industrial set-up. We then look at the CSCW literature on knowledge and expertise sharing, which forms the central backdrop to our

[2] For an overview of what set-up for bending machines entails, see Sect. 4.4.1.

arguments. We follow by examining literature on embodied action, particularly focusing on the methodological aspects of the bodily and spatial aspects of human action that, in our view, make the debate on its relevance to knowledge and expertise sharing interesting. The section on AR technologies discuss how this type of technology can support the recording and visualisation of knowledge embedded in embodied action. We conclude by looking at some of the literature that shows the potential of CPS for knowledge and expertise sharing. We show that, despite the potential identified in the literature, little has been said about how it actually responds to user needs and requirements. This highlights the relevance of the research study herein presented and outlines the research gap that it sets out to address.

4.2.1 Knowledge and Expertise Sharing in Industrial Contexts

Most of the discussion regarding knowledge and expertise sharing in manufacturing and industrial contexts is located in the KM literature. In view of its scale we cannot provide a comprehensive account of that literature here. However, certain parts of it are pertinent to both our own position and arguments previously made in CSCW, so we will briefly examine those elements as background to everything else that follows.

Studies of knowledge and expertise sharing in KM widely acknowledge the importance of knowledge management as a strategic resource for companies, offering potential competitive advantages in the marketplace (Argote & Ingram, 2000; Grant, 1996; Grant & Baden-Fuller, 2004; Spender, 1996a; Watson & Hewett, 2006). However, none of these studies explicitly examine the role of expertise sharing and knowledge transfer in industrial set-up processes. Instead, the focus is recurrently on elaborating models that might assist in understanding key elements of knowledge sharing among workers (Hau et al., 2013; Nonaka, 1994; Nonaka et al., 2000; Nonaka & Toyama, 2003; Spender, 1996a). This has chiefly resulted in material covering either the essential elements of organisational knowledge creation (Nonaka, 1994; Nonaka et al., 2000; Nonaka & Toyama, 2003) or the motivational aspects that might encourage realisation of successful knowledge transfer, such as generation of social capital (Hau et al., 2013; Spender, 1996a), which refers to the "the sum of the actual and potential resources embedded within, available through, and derived from the network of relationships possessed by an individual or social unit" ((Nahapiet & Goshal,

1998), p. 243). Very little of this literature actually deals with knowledge intensive practices and how they might best be shared.

Practices need to be seen as 'ways of doing'; often predicated on knowing just how something is to be done, what organizational and other expectations attend on it, and how it can be recognised for what it is (Schmidt, 2014). Such matters are typically glossed as 'contextuality' but our task here is not to rehearse the point that knowledge is contextual but to show how, in this case, it is. This is particularly important where skills are manifestly 'embodied', relying on a 'feel' for things (see e.g. (Clarke et al., 2003)).

4.2.2 Knowledge and Expertise Sharing in the CSCW Literature

Over the past few decades 'expertise sharing' has become a well-established field within the CSCW literature. Ackerman in particular has focused on this topic, bringing practices and the social setting to the centre of the CSCW discourse about how knowledge and expertise might be shared. Ackerman's work advocates acquiring a detailed empirically-based understanding of the practices associated with knowledge transfer, something that had hitherto been neglected in the literature (Ackerman & Halverson, 2004).

Amongst other things Ackerman's work has touched upon the notion of Organisational Memory (OM). This concept focuses on the storage of organisational knowledge in artefacts such as relational databases (Ackerman, 1998; Randall et al., 1996). For Ackerman, an essential step in OM is documenting an organization's structure to capture features that might be relevant to its 'memory' (Ackerman & Malone, 1990). Related literature discusses possible obstacles to the storage and use of OM. First of all, accessing memory can be troublesome (Randall et al., 1996) and many problems can arise, especially in distributed organizations or where a significant number of people are involved (Ackerman & McDonald, 1996).

However, OM involves more than just data storage. Over time, different systems and interests have evolved, but there has never been final agreement on the exact nature of the support such systems should provide (Bannon & Kuutti, 1996). At a deeper level, the metaphor of 'memory' has also been brought into question. Randall et al. (1996), for instance, argue that the:

> *'organisational memory' metaphor fails to distinguish the kinds of socially situated 'remembering' that might take place in organisational life, and provides few examples of the 'remembering how', remembering who', and 'remembering that' we are interested in ((Randall et al., 1996), p. 29).*

This critique resonates strongly with discussions of the grammatical order of concepts such as 'knowing how' and 'knowing what' outlined above. Thus, we return again to the problematic notions of tacit knowledge and propositional content.

It is worth noting here that, within the knowledge and expertise sharing paradigm, a distinction has been made between two research traditions regarding how knowledge and expertise flows among knowledgeable actors: 'knowledge sharing'; and 'expertise sharing'. These two ideas share many features and overlap in numerous ways but are differentiated by the status of the externalisation of knowledge. While externalisation plays an important role in knowledge sharing, expertise sharing is taken to be less dependent on prior externalisation (Ackerman et al., 2013; Pipek et al., 2012). This resonates with a view that expertise largely corresponds to embodied knowledge (Fitzpatrick, 2003). None of this, however, quite puts to bed the controversy surrounding 'tacit' and 'explicit' knowledge (Nonaka et al., 2000; Polanyi, 1967).

A core criticism here relates to how the concept of tacit knowledge confuses something that was originally a clear practice-oriented consideration. Schmidt (2012) puts forward a very strong argument against the dichotomy between tacit and explicit knowledge, expressing the view that the notion of tacit knowledge is 'a conceptual muddle that mystifies the very concept of practical knowledge' (p. 163). He, as ourselves, favours a focus on practices concerning the actual application of knowledge and their concomitant outcomes. For him, the important issue is to find out who uses which tools and to what level of detail. He also suggests a need for a deeper understanding of 'didactic practices' or 'mutual learning'.

Another way in which the sharing of knowledge and expertise has been differentiated in CSCW research refers to rhetorical focus. According to Ackerman et al. (2013), the first generation of knowledge and expertise sharing research concentrated on the technical characteristics of knowledge management. Thus, the creation of databases containing instructions and standard procedures were considered to be of central importance. However, the neglect of the lack of incentive structures (Hsieh et al., 2011; Orlikowski, 1993; Tausczik & Pennebaker, 2012) and of the social-technical gap created new issues instead.

Following the identification of this social-technical gap, CSCW research invested in illuminating the organisational and social realities of knowledge and expertise sharing. This was deemed to be imperative for the success of KM initiatives (Ackerman & Halverson, 2004), especially local and personal knowledge, which, it was argued, is difficult to capture (Ackerman et al., 2003). This led to a focus on notions such as communities of practice (CoP) (Ackerman et al., 2013; Wenger, 1998) and social capital (Nahapiet & Goshal, 1998), with CoP offering insights regarding the inherent relationship between shared practices and learning opportunities, and social capital assisting with the understanding of the conditions that could foster knowledge and expertise sharing.

Elaboration and discussion of these concepts represented a significant advance in our understanding of knowledge and expertise sharing. However, many questions remain unresolved. Our findings suggest that the missing piece is a focus on the possibility of transmitting non-propositional knowledge through observations of embodied action. To date, no research in CSCW has been addressed to this.

4.2.3 Embodied Action and Knowledge Transfer

The notion of 'embodiment' has a long, and sometimes controversial, history. Research on embodied action has been, inter alia, influenced by Merleau-Ponty's philosophy of embodiment (Crossley, 1995; Gallagher, 2010) and by Wittgensteinian philosophy (Wittgenstein, 1953). For ethnomethodologists and others, however, 'embodied action' is a visible and accountable matter, one where the purposes and rationales that actors bring to their activities can be identified in and through what is said and done (Lynch, 1997). As such, it becomes an empirical matter. In keeping with this perspective, then, we should emphasise here that the fact that knowledge and action are interrelated, and the fact that action is inevitably embodied, is not the point. This has been made abundantly clear by Merleau-Ponty and, before him, Wittgenstein. Our focus is on how, in this particular context of embodiment, knowledge in action is manifest.

We take an interest here, then, in how embodied action 'in practice' can be represented so as to confer information about meaningful behaviour that embodies 'good practice' and that is therefore valuable to share with others. Although research in CSCW has long had an interest in the sequential organisation of 'talk-in-action', Benford et al. (1995) nonetheless noted that 'embodiment often seems to be a neglected issue' (p. 242) in cooperative systems. However, some studies have examined the reciprocal influence of actions and interactions in the virtual and physical worlds (Bowers et al., 1996; Heath et al., 1995; Heath & Luff, 1991). Heath and Luff's (1991) study of video-mediated communication, in particular, identifies the subtle but crucial ways in which the elegant and efficient management of collaborative work is realised through gestures and other visual behaviours.

Others have focused specifically on how embodied action is manifest in the treatment of physical artefacts (Brown & Duguid, 1994; Harrison & Minneman, 1995), for instance: the expression of intention through the movement of physical objects (e.g. repositioning an object in a workspace); the production of physical representations (e.g. writing or drawing on paper); deictic emphasis (e.g. underlining on texts); and the actual use of physical objects (e.g. working with computers, seeing and feeling objects, reading, etc.). According to Kendon (1990), and largely in keeping with Heath and Luff's (op. cit.) insights, embodied action can also be related to other bodies. Significant aspects here include issuing and receiving 'signs' in the form of visual indicators such as co-positioning, specific postures, gestures and facial expressions, not to mention oral and/or acoustic indicators, e.g. speaking and other (external) sounds. If we take these insights on board, it is apparent that supporting these embodied actions will require considerably more than recording propositional content. As we shall see in Sects. 4.2.4 and 4.2.5, although video and similar materials provide some access to our understanding of the 'embodied', we believe that new technologies will afford a richer set of possibilities.

4.2.4 AR, Embodied Action and the Potential for Knowledge and Expertise Sharing

The beginnings of AR technologies date back to the early 1990s (Caudell & Mizell, 1992; Feiner et al., 1993). From their earliest developments, their potential for blending useful information with real world surroundings has already been visible. With the emergence of new technologies like tablets and smartphones, the popularisation of such applications became a real possibility.

Research here has explored two main input and output mechanisms: conventional keyboard/mouse and displays (Gauglitz et al., 2014a); and the use of touchscreen and head-mounted displays associated with voice commands (Fakourfar et al., 2016; Gauglitz et al., 2014b; Zheng et al., 2015). Findings point towards users preferring the later solution. More recent developments have explored the use of gestures as possible interaction mechanisms for AR-based systems (Pollalis et al., 2017), but these are very early developments that need further research.

In terms of applications, much of the focus of existing research has been on mobile AR-applications to support synchronous remote or co-located collaboration. These cases commonly anticipate the in situ creation of content and its distribution among the actors (Henrysson et al., 2005; Langlotz et al., 2010). In keeping with this trend, Gauglitz et al. (Gauglitz et al., 2012, 2014a, 2014b) examine the scope for telecollaboration mediated by AR technology. In their work, sophisticated tracking algorithms have been devised to help remote users to place location-based annotations without marker recognition in the field of view of a local user. On the basis of this, novice users were able to follow instructions from experts, who were remotely placed. Evaluation showed that the proposed technology could aid novice users in following complex instructions quickly and with almost no errors (Gauglitz et al., 2012). Further developments of this application enabled remote users to act independently of the field of view of the local users, whilst still placing location-related annotations. Further results indicated that the proposed solution was both easy to use and of significant help in fulfilling the tasks in the described context (Gauglitz et al., 2014b).

The use of AR in knowledge-intensive environments to support learning processes has also been explored to some extent. A promising feature is fast, unrestricted and context-specific access to information (Klopfer et al., 2005). This type of technology allows for the following affordances, which are likely to be of use in an industrial context as well: (1) exploration of teaching materials from different angles; (2) supporting teaching where no real-world first-hand experience is possible; (3) enhancing collaboration between students and instructors; (4) helping students to control their speed of learning and decide which pathway to choose (Yuen et al., 2011). Findings from the literature also suggest that AR technologies can enhance workers' skills and perception, promising potentially sustainable improvement in things like assembly operations (Ong et al., 2008; Pathomaree & Charoenseang, 2005).

Other areas of application include construction and maintenance work (Lee & Akin, 2011; Wang & Dunston, 2006); inspection work (Chung et al., 1999; Dunston & Shin, 2009); and welding (Park et al., 2007).

Unsurprisingly, embodied action has arisen as an issue in this area. For instance, Fakourfar et al. (2016) highlight the role of deixis in AR systems, i.e. the use of gestures in the course of interaction, noting that deixis facilitates collaboration by allowing for common reference, thus acknowledging the importance of the bodily and spatial aspects of human action. This gives particular substance to the relevance of embodied action to knowledge and expertise sharing. Fakourfar et al. (2016) stress that AR annotation systems need to accommodate deixis, even if this leads to further challenges. Irlitti et al. (2017) comment on how the combined shared space provided by many AR systems allows people to use the same non-verbal cues they use in face-to-face interaction. Tang et al. (2007) explore the role of embodiment in particular detail in the use of mixed presence groupware, i.e. groupware supporting interaction between groups where some people are co-located and others are remote. The authors note that, when such systems do not account for the embodied actions that occur during collaboration, it leads to a phenomenon known as presence disparity[3]. The authors therefore suggest the need to provide mechanisms for embodiment in mixed presence groupware that can offset presence disparity.

Despite Tang et al.'s work (2007) we have found no discussion in the literature regarding how embodied action captured through AR systems can be used for knowledge and expertise sharing. Furthermore, little attention has been paid to scenarios of asynchronous collaboration (Irlitti et al., 2017), which is core to our own research. We set ourselves to address this gap in the literature and to contribute towards reducing it.

4.2.5 CPS and the Lack of Applications for Knowledge Sharing

Current and past research has shown that new technologies, when designed with a view to the social-technical context in which they will be deployed, can potentially support and facilitate numerous different processes[4]. The rise of AR and sensor technologies provides for new opportunities relating to the haptic and thence to a more nuanced understanding of the sequencing of activities. When allied to procedures for examining 'rationale' this offers potential for a richer picture of how knowledge and expertise are constituted and thus shareable. Cyber-physical systems (CPS), as they have come to be termed, have appreciable promise.

[3] Presence disparity occurs when remote collaborators are not sufficiently aware of those who are co-located. According to Tang et al. (2007) it happens because embodied action is key for the processes of: feedback and feedthrough, whereby people perceive themselves and one another; consequential communication, where people grasp information by observing the work of others; and gestures, which correspond to 'bodily movements and postures used for communicative purpose'.

[4] Note that we do not buy into technological determinism but acknowledge technology's assistive potential.

CPS are systems of tightly coupled physical and cyber components that integrate software, hardware, sensors and actuators. The general definition of CPS implies that they can link and manage distributed systems consisting of physical and virtual entities (Lee et al., 2015). Interaction with such systems takes place via a human-machine interface, which can be implemented through conventional PC interfaces, touch screens or even AR-based technologies, as seen in Sect. 4.2.4.

The possible application of sensor technology as a central feature of CPS, is extensive. In an industrial context, for instance, it has the potential to enhance human sensory access by controlling areas that are inaccessible to humans (Chong et al., 2003), and to provide additional information (Gellersen, 2005) of assorted kinds. Hu et al. (2016) and Marseu et al. (2016) discuss three types of sensor/actor relationship: (1) sensor-to-sensor interaction, (2) sensor-to-actor interaction; (3) actor-to-actor interaction. We argue that it is sensible to think of a fourth relationship—i.e. actor-to-sensor interaction—in situations where actors manipulate the outcome of a sensor.

Chen and Tsai (2017) look at the use of sensors in manufacturing and representative cases applying RFID and auto ID. The exact nature of a deployment depends substantially on the dynamics of the physical process and the degree to which it can be automated. At one extreme there is a notion of lean infrastructure without external interfaces. Here, sensors communicate with each other using a knowledge base that contains artefact-based knowledge about the current state of the environment (Strohbach et al., 2004). This requires no human interface because the knowledge base consists exclusively of data and rules for its processing (op. cit.). At the other extreme there is the notion of a CPS being totally supervised by humans. Here a key concern is how data can provide the right information for the right purpose at the right time to the right person (Hu et al., 2016; Lee et al., 2013).

Examples of CPS applications can be found in numerous domains including power electronics, aerospace, defence, energy systems, healthcare, and transportation (Hu et al., 2016). A useful overview of their potential and associated challenges can be found in Park et al. (2012) and in Poovendran et al. (2012).

In an industrial context CPS have been termed CPPS (Cyber Physical Production Systems). CPPS have been proposed as an answer to the progressive mechanisation and networking of machines and sensors, resulting in a flood of data (big data) that can be put to a variety of ends. Under the generic term Industry 4.0[5], this is seen to represent opportunities for entire industries to position themselves competitively for the future (Kagermann et al., 2013). Amongst other things, these systems have been considered for: predictive maintenance, automatic detection and remote diagnostic systems (Chen & Tsai, 2017); the development and operational phases in production (Herterich et al., 2015); and the location of trends and the scope for optimisation (Shi et al., 2011). Other research relates to software development for smart factories (Otto et al., 2014) and CPPS-based modelling (Thramboulidis & Christoulakis, 2016).

In view of CPS's potential for recording the haptics and detailed variables involved in work processes, which can in turn be visualised through AR-based technologies, it is conceivable that such systems may afford a methodological revolution in the gathering and disposition of knowledge, including the knowledge embedded in embodied action. As Wolf (2009) has argued:

> *In the past, we have brought our information to computers in the predigested form of keystrokes and mouse clicks. Cyber physical systems actively engage with the real world in real time and expend real energy. This requires anew understanding of computing as a physical act—a big change for computing (p. 89).*

Engaging with the real world entails knowing a great deal more about how (cooperative) activities and computation might be interwoven. Schirner et al. (2013, p. 36) have referred to this as 'Human in the Loop Cyber Physical Systems' and such an endeavour, as they say, 'poses tremendous challenges'. To begin with there are potential difficulties stemming from the blending of interaction between physical and virtual worlds, resulting in increased interactional and technical complexity. As these are quite novel systems, there is also a need to develop prototyping tools to support the ideation process and user-centred and participatory design (Lee, 2008; Monostori, 2014; Paelke & Röcker, 2015). The main issue, however, concerns the applicability of CPPS and Industry 4.0 concepts to small and medium-sized enterprises (SMEs) (Ludwig et al., 2016). Another noted issue—that goes to the heart of our interest here—is the difficulty of converting the knowledge of skilled workers into program code (Bracht et al., 2011). Further areas of tension concern quality requirements, employee qualification and human-machine cooperation (Ludwig et al., 2016).

[5] Industry 4.0 (I4.0), is a movement that has resulted from the German federal government calling for a 'fourth industrial revolution', inspired by the use of CPSs in manufacturing contexts (Paelke & Röcker, 2015; Wan et al., 2015).

The aforementioned challenges remain subject to further research. In addition to that, it is worth pointing out that CPS-focused research has so far been mainly technology-driven. Thus, there is an evident need for research employing real-world user-centred design approaches to properly understand the requirements present in the social-technical contexts where such applications will be deployed. Furthermore, to the best of our knowledge, no research has yet examined how such systems might support knowledge and expertise sharing by gathering the knowledge embedded in embodied action. This paper is therefore geared towards addressing these two issues.

4.3 Methodology

The findings presented here come from a research project investigating opportunities for the design of technologies to support industrial set-up, a core and time-critical process whose optimisation is directly associated with enhanced productivity and competitiveness (Palanisamy & Siddiqui, 2007; Shingo, 1985). Our research has been conducted under the auspices of the Design Case Study (DCS) methodological framework (Wulf et al., 2011). This approach involves the deployment of ethnographic (and other) approaches to the long-term study of existing practices, socio-technical innovation, the design of new artefacts (where necessary), and study of their long-term appropriation. We report here on the results from the first phase of the DCS cycle, also known as the pre-study. At this stage, we are primarily involved in understanding how existing practices are accomplished and what the possible ramifications for the design of new technology might be.

Ethnographic research, on which much of our early work is predicated, primarily aims at 'making the invisible visible' (Goodwin, 1994) by describing the implicit, social and cultural organisation that characterizes participants' activity (Anderson-Levitt, 2006; Heath & Luff, 1991). Although there may be many versions of what it takes to do this, we draw here on a more or less 'interaction analytic' view, which stresses the sequential organization of activities and the reasoning processes undertaken by those performing them and others who seek to understand them. Through their collaborative work, in this view, actors often engage in activities that are closely linked to each other, without necessarily explicitly discussing their actions (Heath & Luff, 1991). Put simply, mutual understanding is arrived at without recourse to explicit instruction. In keeping with Peter Winch's (Winch, 1964, 1997) position, we see a relationship between the way we understand others and the way we understand ourselves. As he asserts,

'My understanding of 'what I do' ... manifests itself in two ways: in the character of my behaviour itself; and in what I may say about my behaviour.' (Winch, 1997, p. 195). All of us, experienced operatives or ethnographers, may perfectly well understand how we do what we do, but how we describe it to others is a different, and highly contextual, matter.

In the field of 'Workplace Studies' there is a growing interest in various way of representing visible behaviour, conversations and other possible behaviours, for analysis e.g. the interactive organisation of activities in: offices and call centres (Moore et al., 2010; Murphy, 2004; Whalen, 1995; Whalen et al., 2002); operating rooms (Hindmarsh & Pilnick, 2002; Koschmann et al., 2007; Mondada, 2012; Svensson et al., 2007); medical consultations (Beach & LeBaron, 2002; Greatbatch et al., 1993); etc. Much of this work has entailed careful analysis of video material and, again, there is an extensive literature discussing how such material can be made tractable for analysis (Heath et al., 2007; Knoblauch et al., 2008; Mondada, 2006). In our case, because we needed a very close-grained analysis of set-up, we used video recordings and eye-tracking technology.

Specifically, shadowing sessions (Czarniawska, 2007) were performed for the collection of in situ data regarding work practices and social interaction. Supplementary eye-tracking videos provided details of the practical steps during a set-up process[6]. This, together with the transcribed comments of the participants, meant that even complex process steps could be recorded and evaluated in their entirety. Accompanying in-depth interviews (Hermanowicz, 2002) were conducted to enable detailed examination of the practices of knowledge transfer. These were also audio recorded and transcribed. A total of 14 shadowing sessions with accompanying eye-tracking recordings were performed. With the help of fieldwork notes, the interactions were documented before, during and after the sessions. There were 24 interviews of between 45 and 120 minutes. The transcription process used the 'intelligent verbatim' method.

In total, 24 workers from 4 different SMEs between the ages of 20 and 60 participated. 7 belonged to Company A, 13 to Company B, 2 to Company C,

[6] We used an eye-tracking technology to record actual set-up sessions in the course of ordinary workdays of our participants. By capturing the exact gaze of participants, we were able to get a better understanding of the operations they carried out. Eye-tracker recordings gave us both a better view of the parts being exchanged and the actions involved in the exchange. We also got participants to use a Think Aloud protocol (Nielsen, 1993, p. 195) during the session, so as to gather information about their reasoning as the set-up progressed. Analysis of this data gave us a deeper understanding of what industrial set-up entails and underpinned heat-map analysis of the tools and parts the participants were focusing on.

and 2 to Company D. All companies were medium-sized and produced cold-forming parts. The interviewees held various positions in the company, came from different educational backgrounds and had different job tenures (see Table 4.1). We wanted to cover a diversity of profiles to gather a more nuanced account of the investigated themes. This is common practice in studies focusing on the design of technologies for a wide spectrum of end-users (Braun & Clarke, 2012; Sharp et al., 2006).

Table 4.1 P: Participants in the empirical study, I: Interviews, E/S: Eye-Tracking/Shadowing. Education (u/s): Education (unspecialized/specialized)

P	I	E/S	Role	Company	Education (u/s)	Job tenure
P1	X		Foreman	A	graduated (u)	> 10 years
P2	X		Designer	A	Master school (s)	> 10 years
P3	X		Foreman	A	Master school (s)	> 10 years
P4	X		Production engineer	A	graduated (u)	> 10 years
P5	X	X	Machine setter	A	Master school (s)	< 10 years
P6	X	X	Machine setter	A	Master school (s)	< 10 years
P7	X	XX	Machine setter	A	Apprenticeship (s)	> 10 years
P8	X	XX	Machine setter	B	Apprenticeship (s)	< 10 years
P9	X	X	Machine setter	B	Apprenticeship (s)	< 10 years
P10	X	X	Machine setter	B	Apprenticeship (u)	< 10 years
P11	X	XX	Machine setter	B	Apprenticeship (u)	< 10 years
P12	X		Process owner	B	Apprenticeship (s)	< 10 years
P13	X		Foreman	B	Master school (s)	> 10 years
P14	X		Technical salesman	B	graduated (s)	> 10 years
P15	X		Technical salesman	B	Master school (s)	> 10 years
P16	X		Construction engineer	B	graduated (s)	> 10 years
P17	X		Technical salesman	B	graduated (s)	> 10 years
P18	X		Production salesman	B	graduated (s)	> 10 years
P19	X		Quality engineer	B	graduated (s)	< 10 years
P20	X	X	Foreman	B	Apprenticeship (s)	> 10 years
P21	X	X	Machine setter	C	Apprenticeship (s)	< 10 years
P22	X	X	Foreman	C	Master school (s)	> 10 years
P23	X	X	Machine setter	D	Apprenticeship (s)	< 10 years
P24	X		Production engineer	D	graduated (s)	> 10 years

The interview transcripts, field notes and eye-tracking recordings were subjected to Thematic Analysis, which entails a set of well-established steps involving open coding of the media excerpts, systematic revision of the coded segments, and identification of code-families and their relationships (Braun & Clarke, 2012). This enables elaboration of a deep understanding of the explored contexts and/or phenomenon (Ayres, 2016; Braun & Clarke, 2012; Gibson & Brown, 2009). The findings from different data artefacts were triangulated to ensure their trustworthiness, a quality criterion for qualitative research (Bryman, 2008). Below, we present the themes uncovered in this analysis. Description of the workflow presented in Sect. 4.4 is derived entirely from our field notes. The findings stemming from it have been checked against the interview and eye-tracking data, which are occasionally presented through the section in the format of quotes or vignettes so as to ground the findings in concrete detail. The materials presented in Sect. 4.5 are mainly based on the interviews carried out after the shadowing sessions, which were informed by our observations during the shadowing sessions themselves.

4.4 The Relevance of Knowledge in Industrial Set-up

Industrial set-up is a process involving a set of preparatory actions on a machine or a tool prior to the start of a production cycle. It happens between the end of the serial production of one article and the start of the serial production of a different one. This is a core and time-critical process from an economic perspective: without set-up there might be no production (Van Goubergen & Van Landeghem, 2002; Voigt et al., 2016).

Set-up can be carried out by a single machine operator (as it was in Companies A, C and D)—or by two or more operators (Company B). The latter usually pertains to larger machines. For this paper we decided to focus on single machine operator set-up as we observed that issues of knowledge and expertise sharing, which are our central interest, are more salient in this set-up scenario.

Set-up, we quickly discovered, is a knowledge-intensive process, relying mainly on experiential—or experience-based—knowledge. The literature indicates that such knowledge can be both highly personal and very difficult to capture and share (Matthew & Sternberg, 2009; Nonaka et al., 2000; Randall et al., 1996).

4.4.1 Industrial Set-up for Rotary Bending

Rotary bending is a manufacturing process involving the use of specialised press machines that can be customised for the production of different articles through the set-up of a complex tool, known as the rotary draw (see Fig. 4.1). The process usually unfolds as follows: first of all, a semi-finished product is fixed between an outer and inner clamp die. Subsequently the semi-finished product is curved by rotation around the centre point of a bend die, with a pressure die stabilizing the straight section of the pipe in front of the actual forming area. A collet guides and pushes the pipe in the forming direction, whilst a wiper die reduces the formation of wrinkles on the inside of the tube and a mandrel avoids necking of the tube by supporting its inner stability (VDI 3430, 2014).

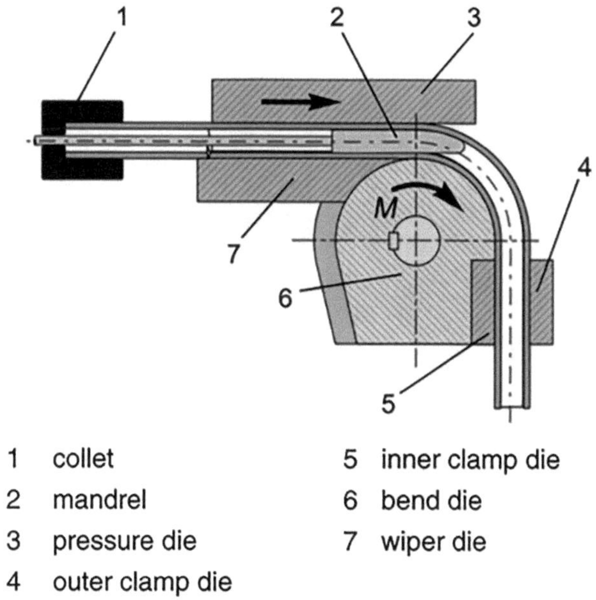

Fig. 4.1 Tool components for rotary draw bending processes (VDI 3430, 2014)

The set-up of the rotary draw involves the assembly and disassembly of its several components on the machine (see Fig. 4.1 and Fig. 4.2) and demands the use of screwdrivers, clamps and several other hand tools, which supports the manual procedures carried out by a machine operator, also known as setter or installer. Logistical and other preparatory activities such as transporting the components and the tools for their assembly and disassembly, also form part of the set-up, as we will see. In addition to that there are the positioning, movement and adjustment of the individual tool components through the configuration parameters in the control unit of the machine. Through these actions installers attempt to reach the point where all the tool components perform their functions satisfactorily, so to delivery of a curved tube—e.g. a chair leg—without any visible problems—e.g. wrinkles or cracks. This process is formally considered to be a tool-dependent cold-forming process.

Fig. 4.2 Mounted tools on a bending machine

4.4.2 Talking Planning and Logistics

Set-up usually starts with procedures based on a production order. This provides, amongst other things, details about the item to be produced and the machine to be used in its production. Using the article number, workers compile a list detailing the set of tools and machine parts required for the set-up. This list is based on a list of parts for bending a particular product associated with each machine. However, these lists are not always up-to-date. Moreover, workers could not say whether this information exists for every single product they work on:

> *We have created this set-up data once for the MEWAG machine, so that people know which tool belongs to which type and which place. I then supplemented it with this type. In the production area we have a corresponding list - at least on the electric machine. I also did it for the hydraulic machine. (P2, Interview)*

Here we have an issue of available knowledge. The quote suggests that there has been an attempt at sharing knowledge about which parts could be fitted to a particular machine, although it became clear that this initiative has not been sustained. Our data suggests that this list needs to be a living document, because new machine parts are designed from time to time to replace old ones and optimise the bending process. Given the unreliability of the data in the list, machine operators have to draw on experiential knowledge to elaborate the list effectively.

> *The main problem is that no tools are complete. We have to assemble a tool from two or three other tools. [...] The setters do not have that much experience. This is very complicated for the setters, so they have a lot to think about. You have to think about which tool needs to be adapted. And you lose a lot of time. (P11, Interview)*

As they continue with compiling the list, the location of the individual tools and machine parts becomes highly relevant. We uncovered recurrent problems concerning the storage of tools and parts. Most importantly, identifying the appropriate parts for the set-up is usually a non-trivial task. Some parts are very alike and it is difficult to tell them apart. Therefore, knowing what is necessary for the set-up, knowing where to find the relevant parts and knowing that a part is really what is necessary for the set-up have been identified as some types of local knowledge needed for a successful set-up process. Having this knowledge allows workers to quickly assemble the components they need to get underway, so it impacts positively on the time performance of the workers.

4.4.3 Assembly and Disassembly Cycles

Once the list is ready, workers physically gather the tools and parts. These are usually collected in a trolley and brought to the machine. Our observations showed that sometimes the required tools or parts are already in use. It was very common for machine parts to already be mounted on the machine to be used for the production order in question, or else on other machines that were currently in the middle of a production cycle. This causes logistical issues. Yet again, knowing that and knowing where come strongly to the fore. Devising ways to give this information in advance would save workers valuable time and contribute towards the efficiency of the set-up operation.

Once at the machine, workers make room for the set-up. Sometimes they have to slightly adjust a machine, so that they can access all the points where parts must be removed from or attached to. This particularly relates to local knowledge and embodied practices.

After adjusting the position of the machine, workers proceed to disassembling its current configuration, so the new one can be assembled instead. Sometimes parts already in the machine can stay where they are, so workers do not always need to remove all components from a configuration to start another. It varies as to whether they just need to swap two parts or remove several before a new one can be fitted. The decision on which parts should stay and which ones should be removed is based on experiential knowledge, which differs from worker to worker. However, our informants acknowledged that there are always improvements that can be made:

> *I've been in the company for a little over two years now, and I've come to the point where I master the manufacturing cell F-03 very well. And there are still tricks that my colleagues who are here for 15 to 20 years can tell me - and they do. I just have to ask questions. This also applies to the operators. The operator also knows tricks that can be passed on through experience and questions (P5, Interview)*

Experience, then, teaches workers the shortcuts that can be taken and how the set-up can be optimised. Here, in contrast with the previous situations, we move towards an issue of expertise sharing: sharing the 'knowledge of how' to best accomplish a task. Evidently, sharing this kind of expertise could help novice workers enormously:

> *Only the practice keeps you going - try, try, try. The experienced are the best employees. They understand everything faster. Anyone can study theoretically, but in practice, that's something completely different. (P8, Interview)*

In fact, we sometimes saw workers become so dependent on the knowledge of other colleagues, that production could be delayed if they were not around. P6 reported this as follows: *"The [set-up] order is only determined on the basis of experience...The segment leader has been familiar with this for a long time. When he is on vacation there are always problems." (P6, Interview)*

Acknowledging the importance of such knowledge, some participants raised the issue of making available documentation that could enhance productivity:

> *Productivity would certainly be present if a set-up guide were already available and if this guide could then be presented through a tool. The tool should again be fitted relatively easily for the single machine. So I can say then: 'I only mount the tools, but the whole guide is already there'. For my sake, on an iPad or whatever. (P4, Interview)*

Assembly is guided by the list of defined tools and parts collated before the actual set-up starts. However, there are no guidelines regarding what to disassemble in the course of the set-up. Parts must be mounted in the exact location on the marked machine axis. These steps are also deeply rooted in experiential knowledge:

> *The hole must be in the middle, the bow must be clean, there should be no scratches on it. And now it has to fit into the template here. Distances at the beginning and at the end must lie within the markings. He has to fit in all the cams. The end must not be farther back than the first stroke and the white stroke here is the longest run. So, now I can start the system and continue to watch the first 15 minutes, because there is still something to do. (P5, Eye Tracking)*

The information needed can only be found in manuals with difficulty. The knowledge is embedded in a workers' routine and so needs to be understood and deployed contextually. Nevertheless, these tricks and tips, subtle though they are, are deeply relevant to efficient and speedy set-up.

The alternation between assembly and disassembly of parts continues until all new parts are fitted in the machine. It is important to note that sometimes assembly is interrupted for other reasons. Sometimes workers forget to include important tools or parts in their initial list. Sometimes set-up is interrupted by colleagues coming over to ask for help with the set-up of other machines. Sometimes, managers tell workers to stop a task and attend to more pressing matters.

Operatives argued that recording interruptions arising from problems, breakdowns, solutions and work-arounds, was important for others to know about as a source of learning:

> *That would perhaps help us for those who do not yet have a comprehensive technical background. You will then be introduced to the matter faster. You might as well take a quick look 'Ah, there's the problem I've caused. (P1, Interviews)*

Industrial set-up currently can, and often does, lack systematization, and is mainly based on the experience and 'know how' of a respective machine operator. Disassembled tools are temporarily stored, often chaotically, in a trolley and only properly put away after completion of the final adjustment of the machine program.

4.4.4 Programming and Adjusting

Once the assembly of the parts ends, it is time to work on the machine program, which is already stored in the machine controller. Workers usually interact with the program through a touch screen display or through a remote control-like device. This part of the set-up process is highly dynamic, characterized by constantly changing settings on the machine. In the age of computer numeric controlled (CNC) machines, corrections are made primarily in the program code of the controller. In cases of a control deviation between the bending component (actual value) and the drawing (set value), operators act as regulators by independently determining the manipulated variable from the observations made—see Fig. 4.3.

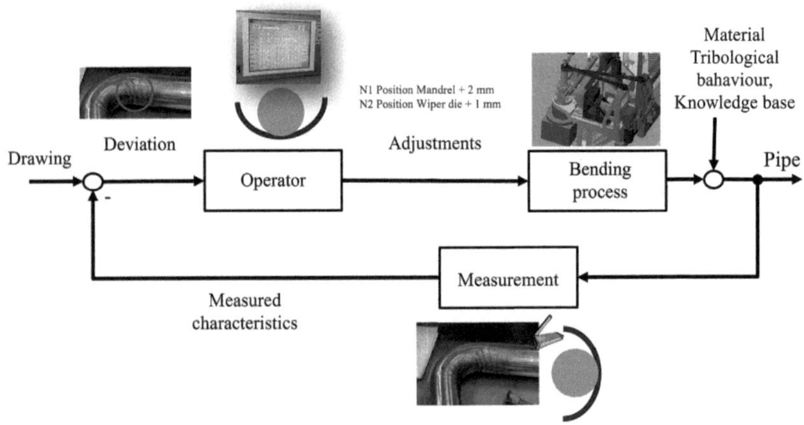

Fig. 4.3 A man in the loop of a set-up process

The dynamics of the set-up process are largely caused by the existence of fluctuating disturbance variables. These can be due to creeping influences, such as wear, or to ad-hoc matters such as changing material parameters. Machine operators rely heavily on their experience for the correct interpretation of measured values and visible phenomena. Their efforts lead to a definitive set of parameters, which are stored as correction values within the CNC program. At this point, it is possible to see a discrepancy between the inaccurate information—on the basis of which the corrections take place—and the need for very precise settings in the CNC code. Figure 4.3 shows a correction of pleats on an inner pipe bend. The machine operator recognizes this issue and defines the manipulated variable. In this case, the operator changed the position of one machine axis by 2 mm and the position of another axis by 1 mm. However, he was unable to explain the cause of the problem. Corrective actions like this can take up half of the actual set-up time and even lead to interruptions during actual production.

Parameter adjustments are usually made according to information acquired through practice:

> *The segment leaders in the production area are so fit that they can already see by the colour of the material whether it's a harder or a softer material. And they also know that they, then, have to change the parameters. Either way, they notice it. I do not want to believe that at first, but I see a difference in the meantime as well. I do not know if it's hard or soft, but I recognize the difference. (P2, Interview)*

Our findings suggest there are so many parameters they could not possible be kept in a physical catalogue for operator support during set-up:

> *There are many parameters involved. For example, if I have materials between 530 Newton and 640 you'll hear it yourself, there's a huge difference. And watch every single step. How do I want to hold on to that? If you could program a machine like that, others would have done it before. This is still missing in bending technology. This was a three-roller bending machine that should automatically deliver. It was then determined that the machine has done everything. She bent the one part for hours. It did not work. We did not get the radius it should have. That did not work. Due to the material deviations. (P5, Interview)*

Once the parameters are set, workers go on to test the results of the set-up. They typically take one of the pipes that will go into the production cycle, bend it and carry out a visual quality test. First, they look for visible irregularities in the bend, e.g. cracks or wrinkles. Then, they put it in a testing mould to make sure that the resulting angles are correct. If the pipe is of sufficient quality, the set-up process is finished at this point. However, we found that workers seldom get it right at the first attempt. A series of adjustments are usually necessary, through which workers deal with the correction of dimensional deviations. The machine interface is used to change the paths and angles of the bending axes. To facilitate this, the exact position of the deviation is detected first of all. With this information in hand, the relevant line of the machine code is selected and a correction parameter is entered. After this, the bending process is re-run and the resulting tube is again subjected to a gauge test. This process is usually repeated several times because the parameters influence each other in unpredictable ways such that, following a correction, various other portions of the tube have to be corrected as well. This creates an overhead of lost time and wastes material, as a new pipe has to be used for each test.

All of this is heavily based on the installer's experience and related information is not kept anywhere in the company. In one eye-tracking session we observed P5 working on adjustments to the machine programme. He went to the correction window, and changed the value from 1.3 mm to 0.8 mm. This decision was taken after a brief glance at the component in the gauge. As one of the values had been adjusted by 0.5 mm, this cascaded to the others values as well.

In another eye-tracking session, P6 recounted his reasoning as follows:

> *I then go to my programs. Here are my individual bending programs [a multi-step program]. I now consider which individual bends I adjust. This is the last* 90 ° *bend of the part. This is the process number 7. It is now over bended to 0.5 °. But I want to turn it up a bit, that is, we'll set it to 0°. I save my change again and drive the machine into the basic position so that it picks up the data. And now turn the next part. (P6, Eye-Tracking)*

There is evidently a great deal of 'know how' embedded in the observed assembly/disassembly cycles, knowledge that is not easily recovered. We noted that less experienced workers needed more time to make adjustments and wasted more production material on tests. So, assuming it can be accomplished without incurring other costs, the sharing of this embodied knowledge and expertise amongst workers would be of evident value to this kind of enterprise.

4.5 Knowledge and Expertise Sharing Challenges in Industrial Set-up

Fostering knowledge and expertise sharing is widely acknowledged to be an investment in a company's competitiveness (Ackerman, 1998; Gupta et al., 2000; Schmidt, 2012). During our investigation, participants consistently referred to the importance of promoting practices of knowledge and expertise sharing. However, there are clearly several obstacles to accomplishing this.

4.5.1 Strategies for Knowledge and Expertise Sharing and Associated Difficulties

The primary source for knowledge and expertise sharing in our companies was training. According to our informants, new workers have to undertake training before being allowed to engage in set-up. Segment leaders, who usually have 2 years plus of experience in the company, are responsible for introducing new workers to the work procedures and for providing them with all the necessary information:

> *[Training] courses take place here in the house. Either internally or externally. Then there will be a briefing by a colleague - usually a segment leader colleague who is already experienced in the field. They also do it very well. I can only confirm that. The information that has to flow also flows. But there are many questions that need to be considered. And these questions only come up in the course of time. (P5, Interview)*

Although training is necessary, it is not sufficient. Set-up, especially programming of the machine, is a very complex process. You have to know a lot already. *"There is a lot for newcomers to learn!" (P6, Eye-Tracking).* Unsurprisingly the didactic practices of segment leaders are based on demonstrating the actual job, not on verbal or written descriptions alone:

> *When I teach [people], I take [them] with me when I rebuild. I tell them, 'Here are the parts. If you cannot figure out which parts go where, you have a list here. 'Then I do the set-up with them four to five times. At some point, I say, 'I showed you this a couple of times. Now you do it yourself '. Of course I'll go with you, check and see how he gets started. What does he do? That actually works pretty well. This worked very well for a colleague. (P7, interview)*

But even the most willing trainers can forget to mention relevant issues. One possible reason for this is the fact that there are things that are so ingrained in their practice, they tend to perform them automatically, without reflection:

> *People are different. The only problem I see is that... There are people who teach others by showing exactly how the process happens. [They will say:] 'Then you have to pay attention here. Then you have to pay attention to that'. But there are also people who take some things for granted, because they are 20 years and then forget to mention... Even minor details. And when a situation comes up [they will say] 'Oh I forgot to tell you that ...'. If you ask a colleague who has been here for some time, he also likes to pass on his knowledge. (P5, Interview)*

An alternative to personal transmission of knowledge and expertise is learning through existing documentation, such as the photo album discussed by Schmidt (2012). However, in this case, documentation has proven to be less than adequate, partly because there is little motivation to keep it properly updated. Instead, knowledge tends to be concentrated 'in the heads' of very few actors:

> *[...], there is just a lack of documentation so this mainly remains 'in the head' knowledge of the individual employees. If today three employees leave the company and tomorrow three new ones are hired, then a massive problem arises. (P4, Interview)*

Those documents that do exist concerning set-up are usually very abstract. Often, the manuals in question are machine-, rather than product-specific. This means, they usually describe how different parts can be attached to the machine, but do not specify which parts should be used for which product, making the lives of new workers hard. These documents are also often outdated:

> *The document [...] was prepared at the very beginning, which was for the first few programs of the first articles. And then, with the time when we got more and more articles, the whole [set-up process] has been changed. The documentation has remained as it has been originally conceived. (P8, Interview)*

As a result, existing documentation ends up not being used, and the main source of knowledge and expertise sharing remains personal exchange.

4.5.2 The Burden of Knowledge Sharing and Ways to Relieve it

Finding the right person with the requisite expertise is not a particular problem in this context. These are, after all, SMEs and a relatively small group of people is engaged in set-up during each shift. Furthermore, segment leaders serve as a reference point, so workers do not have to worry too much about finding the right person to go and ask for help. There are also colleagues that can help with different types of problem, each one within their own area of expertise:

> *We always have a segment leader, who is experienced and summoned for help by anyone. So, you are not alone. Then there are the electricians, who can help with technical problems on the robot or set-up problems with the robots. They are always available and know the important information. These men are very well trained. Of course, they are much better at programming robots than I am, at least in part. Everyone*

has their own area of expertise and they are divided according to the shifts. (P5, Interview)

There are, unsurprisingly, often tensions involved in having to repeat some sorts of information: 'Now we have to tell the new people the written instructions again and again. This is complicated' (P11, Interview). This includes the maintenance of tools and machines and, in the case of segment leaders, the need to carry out several set-up operations. Such repeated explanations can get in the way of the primary task. Therefore, capturing the knowledge held by these people in ways that it can be easily shared and understood is key for effective knowledge and expertise sharing. This is a question of devising the appropriate 'boundary objects' to transit between shifts, work groups and contingencies (Ackerman et al., 2013). Some kind of documentation is generally perceived as necessary:

As I said, standardized set-up documentation would be important. If there is anything that must be highlighted, if it were recorded, were documented, it would be easier for everyone [to get to know about it]. That is missing completely. That's missing on all machines. (P1, Interview)

Nevertheless, a theme that strongly emerged from our data is the usefulness of building up a common information space (CIS) as problems happen:

Once an error occurs, it must be written down with a solution. If I write an error and the path that led to it, it may be that the segment leader comes who has been here for 25 years and says 'that is a piece of cake. Was totally wrong. We have to go this way'. You have to weigh that up afterwards. But you actually have to write something down to see how things are. So, everyone remembers about it and makes things differently. (P6, Interview)

If we see this problem as twofold—documenting what knowledge might be shared and identifying the circumstances in which it needs to be shared—this resonates with known approaches to expertise sharing, such Answer Garden (AG) and its sequel Answer Garden 2 (AG2). Here the issue of OM was addressed by providing an architecture to support the retrieval of recorded knowledge and to provide access to appropriate individuals (Ackerman & Malone, 1990; Ackerman & McDonald, 1996). However, time constraints and 'ownership' issues often over-step such solutions.

4.5.3 Technological Challenges

Our analysis suggests that there are possibilities for CPS to support knowledge and expertise sharing. One possible direction, according to the participants, is support for recording the practical steps undertaken during the set-up process. One participant also said:

> *[Knowledge that we cannot write down directly] is very important for the set-up. This knowledge you learn during the practice. So, the new employees are watching and you do not need to write anything for the bending machine because the programs are there, all programs are stored on the computer. You choose the right program, what you should produce, otherwise you do not need to know anything. You have to be careful to select the right program, because if the pipe length is 500 [mm], you cannot choose 1000 [mm] as length. Of course, if the length is 1050 [mm] and the program chooses only for the length 1000 [mm] then there are problems at the end of the bending process. (P10, Interview)*

So, local knowledge embedded in practice is very important for set-up operations. There is an expectation that, if this type of knowledge is recorded and shared, the costs of training will reduce and efficiency and effectiveness increase. However, it is not just a matter of recording the data, but also of showing it in appropriate formats. For instance, one of the participants raised the issue of data visualization:

> *The central thing is the material characteristics. Nowadays, a simulation is still not able to say that 'you have to adjust the axis by 2mm'. Simulation is not able to say that. But the simulation can clarify physical connections to me. It would be desirable that it could say 'The physical relationships in this bending machine are so.' If I can operate it, I am automatically at that level and can represent the physical connections. (P5, Interview)*

Again, the central issue here is that embodiment is not easy to represent in terms of propositional content. As we have seen, video provides a way forward for ethnographers interested in work practices, but new technology affords a more sophisticated set of possibilities that might provide for the more specific needs of inexperienced practitioners. The sheer delicacy of alignments, such that, for instance, adding or deducting 1 or 2 mm from a particular configuration makes a difference to the adequacy of results, seems to indicate the use of sensor technology as a possibility. We will elaborate on this in our discussion but, to conclude, we present the view of another participant regarding how technologies that could record and share 'know how', 'know that' and many other types of knowing embodied in action could revolutionize the sharing of knowledge and expertise:

If you had a novice now, where you would say 'I'm not showing you how to do that. You do it!'. For example, you have a small tool trolley, there is an iPad on it and what you have filmed with me, that you would do it in a more extensive or longer form, where people see [he looks away and takes some the tools] and knows exactly 'I have to take this' or 'put it into the box' and if I build this in the order... every idiot must be able to do it, to accomplish at least a set-up. Changing values and parameters is again a matter of experience. If it's purely about the set-up process of the machine, that one can say 'I'm watching. I have to do it that way. I have to make that up. Or 'that has to be done'. But that is for the absolute novices. (P7, Debriefing interview after first eye-tracking session)

4.6 Design Implications

The findings above indicate the sheer complexity of the knowledge and expertise sharing problem in the explored context. As we have seen, there is a fairly substantial literature dealing with the possible application of CPS across various domains. Few, if any of the studies in this literature, however, base their design on the kinds of detailed qualitative study we have engaged in, nor have they dealt with the possible application of such systems to set-up work. Below, we suggest ways in which studies of this kind can inform the configuration of CPS in industrial settings of the kind we have examined—settings that are still underrepresented in the CSCW and HCI literature. At present, we would not claim to be at a stage where we can present a worked-through design solution to the knowledge and expertise sharing issues we raise above, but this is not in any case our intent. Our point, rather, is that solutions to problems of this kind require us to know exactly what kind of problem we are confronted with, and why it matters. We have shown how, for reasons of both time and money, it does. CPS, if they are to constitute a potential solution, need to offer more than the presentation of simple information or instructions. These are already amply visible in the literature (Klinker et al., 2001; Pathomaree & Charoenseang, 2005; Wang et al., 2014; Zhu et al., 2014), but do not seem to be sufficient, according to our findings, to enable expertise sharing in the level of detail required by an operator. The point, we feel, is to convey 'know how' as well, with the aim, for instance, of enabling an installer to have adequate competence when dealing with tolerances of one or two millimetres. Moreover, as instructors may themselves be heavily engaged in the day-to-day business of production, any solution will need to recognise that there is a time-critical element. Thus, the technology must be able, during the set-up process, to provide adequate, relevant and timely information in sufficient detail that it can reliably inform operatives of the subtle and complex manoeuvrings that may be necessary in a set-up process. It should also constitute a repository that can be used both for the acquisition of information and as a basis for training, insofar as some of the knowledge is of a propositional kind. At present this kind of information is: 1) not always up-to-date and well maintained; and 2) not always easy to find.

There is an established literature that points to how apprenticeship models encompass features of 'learning by doing' (Lave & Wenger, 1991). Thus, over time, operatives do learn the manipulations necessary for effective working by copying others. However, this requires a physically collocated workspace (Tang et al., 2007). The set-up process described above often does not allow for learning of this kind to take place, because the necessary experience and expertise is not always immediately available.

Whatever system is envisaged, then, it has to cope with some evident challenges. In our view AR technology, in combination with appropriate sensory input values and, where appropriate, codified information about tooling and other material properties, can help overcome many of the difficulties we have discussed. Any effective technology, here, will have to deal with the following three main issues:

- *Design for timeliness.* As we have seen, expertise is relatively scarce in this environment and error is costly. Any solution will therefore need to deliver relevant information in a timely way with minimal prospect of error. Some of the relevant information will be codifiable (e.g. parts and machinery that are pertinent to the task) but the repository for this information will have to be easily maintained, up-to-date and capable of providing contingent information as and when required. In other instances, seeing an experienced operative do the assembly will be the best solution available. Here, providing a granularity that reflects the subtle judgements made by experienced operatives is best done by managing sensory data input.
- *Design for appropriate visualisation.* Capturing and reproducing instructive material still requires decisions about the position from which the process will be observed. Findings from the literature shows that capture in this case is best done from the first-person perspective, although further research is still necessary (Fussell et al., 2003). Head-mounted applications are especially well-suited to the set-up of new parts and materials, as they allow workers to use the two hands. In these situations, every move may contain essential information. Capturing complex set-up instructions in a stepwise fashion using short video sequences from a first-person perspective can offer effective support for the visualisation of particular actions that need to be carried out.

- *Design for the contextual use of sensors.* Physical processes need to be rendered by measuring potentially tiny deviations from target processes. Capturing small tolerances through sensory data input is a promising solution, especially since it has the potential to reduce error. Sensors can reduce uncertainty through workspace monitoring, thus increasing the effectiveness of instruction by preventing errors and reducing stress (Miller & Swain, 1987).

Just one option that might meet the above requirements, and that we are currently exploring as an early prototype, would be a set-up editor developed for the Microsoft HoloLens. The basic functions of this device can be subdivided into two areas in a way that might allow for effective knowledge and expertise sharing.

The first area would comprise a recording module. This would enable an expert to record and demonstrate the set-up process with short video clips and other media such as photos and voice annotation. As the HoloLens is a head-mounted device with a built-in camera, this would allow for the capture of embodied actions from the perspective of the installer. The AR aspects of the technology can then be drawn on to ensure appropriate contextualization of the information by showing the exact installation location on the real machine (see Fig. 4.4).

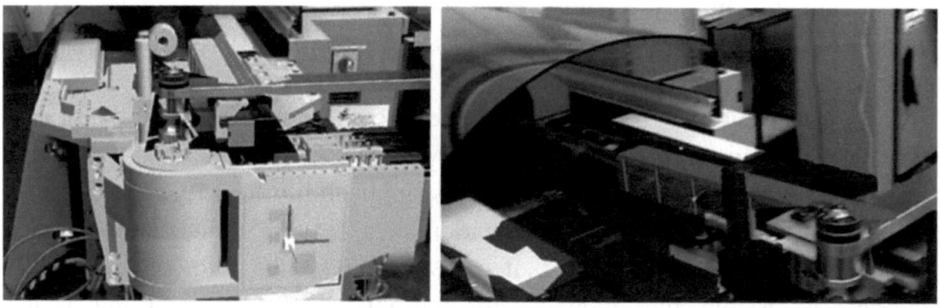

Fig. 4.4 Contextualization of set-up instructions by marking the assembly sites on a bending machine—e.g. the green, blue and red axes on the left picture and the red area on the right picture

The second area would include a playback module where the set-up instructions could be displayed in a context-specific way and enriched with sensor values. For this, the installation location can be marked (see Fig. 4.5) and a video sequence of the set-up instructions displayed. The installer could then independently navigate step-by-step through all of the set-up instructions whilst maintaining access to all the required assembly components and assembly aids.

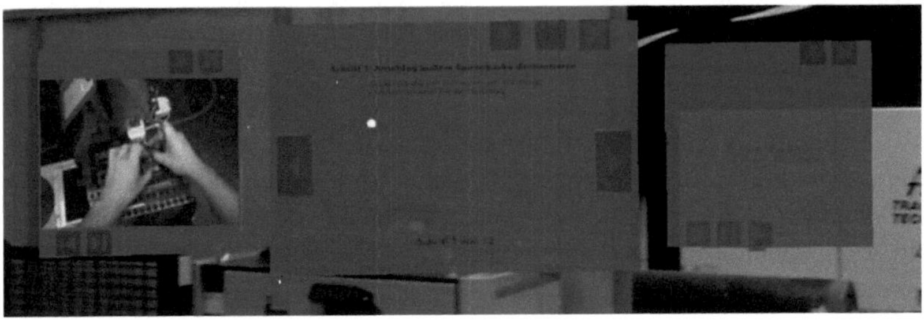

Fig. 4.5 First prototype of the playback module running set-up instructions

Sensors can be used here in two critical ways. First of all, position sensors can be used to measure the displacement of tools and certain machine components. The results can then provide an indication of the outcome of different actions, contributing to a deeper understanding of the process. Mistakes due to the incorrect mounting of tools, or even the presence of a wrong tool, can be detected by the sensors. To do this, contactless measurement can be accomplished by means of a 3D camera. This can reliably detect dimensional deviations independent of light and dirt conditions (in the left-hand picture in Fig. 4.2 an installed camera can be seen at the top of the picture).

As suggested by related work, AR is likely to make formal training less necessary and can be used directly on the job (Neumann & Majoros, 1998), although there is still a need to engender trust in its applicability to actual activities. In relation to this, AR can locate information in a specific milieu and imbue it with specific contextual relevance. It can also render the information quickly and easily accessible, thus reducing any inhibition threshold regarding its use (Neumann & Majoros, 1998). Nonetheless, certain situational characteristics will remain absent. These include environmental conditions. However, in the described deployment these can be detected by sensors, for instance, the location of the tools on the machine or the bending of the machine under a load. The trustability of the provided visualization delivered by the use of sensors and description of the situational parameters in conjunction with the AR technology, can result in a comprehensive artificial collocated workspace. Together these elements can make it possible to integrate embodied action into the expertise sharing process for set-up tasks.

4.7 Discussion

The data presented across Sects. 4.4 and 4.5 shows that, when it comes to industrial set-up, several kinds of knowledge come into play. In so doing, the dichotomy found in much of the KM literature between 'tacit' and 'explicit' knowledge, or 'know how' and 'know that', does not suffice to understand the practices involved. As Schmidt before us has argued (2012), there are not just two categories of knowledge, but 'knowledge of different things' (ibid., p. 214), which involve not only 'knowing how' and 'knowing that', as proposed by Ryle, but also 'knowing what', 'knowing who', 'knowing where', 'knowing when', and so on. Thus, in this paper we have sought to decompose 'knowing how' and 'knowing that' into finer elements, with concomitant proposals concerning technological support for sharing.

Our findings suggest that practiced practitioners are so embedded in their routine work that what seems rather mysterious to outsiders is, for them, utterly mundane and obvious. As a result, some things get taken for granted, which is likely to impact negatively on knowledge and expertise sharing. This calls for solutions that allow 'taken for granted' local knowledge to be shared with others. Furthermore, since much of this knowledge is embedded in reasoning that is manifest in closely sequenced activities, it would seem that methods for displaying those sequences in similar detail are appropriate. At the same time, the critical point here is to identify what explanations are being used, what purpose they have, why it is worth articulating them, and how they can best be represented. We have suggested above that the problem for inexperienced workers is akin to the problem faced by ethnographers in their need to capture the detailed and meaningful aspects of work practice. In the last few years there have been notable improvements in, inter alia, video technology that have provided for increased granularity and better rendering of the subtle and sometimes overlooked features of collaborative work. It remains the case, however, that some aspects, especially haptic and judgemental elements, are still difficult to recover. In addition, unlike ethnographers, operatives and the organization as a whole face significant overhead in terms of time taken, errors made and resulting costs.

Put simply, we are dealing yet again with a practical problem concerning how workers can actually share expertise that is embedded in what they do, and that it is extremely difficult for them to know how to explain when confronted with someone who is not equally invested in such practices. This is only compounded by having to provide explanations as they 'go along'. Explaining is a secondary task, not a primary one. So then, new technology might, with maturity, provide opportunities to dissolve many of these issues related to the distinction between 'know how' and 'know that', whilst also affording other kinds of knowledge sharing.

There are two general issues confronting us. The first one is that this knowledge is local, situated and used in contingent situations. Thus, it only exists within the situations in which one tries to explicate it. The second concerns recipient design (Sacks, 1995). Here knowledge and expertise sharing depend on the relationship between the actors involved. Our data provides evidence that the level of detail or the willingness to share a particular piece of knowledge changes according to the recipient.

To the best of our knowledge, these two issues are usually handled separately in the literature, whilst our data suggests that they are actually interlocked and ramify from one to the other in various subtle ways. Our findings suggest that we need to find mechanisms to overcome the various overheads entailed, so that knowledge and expertise sharing can effectively occur. This is becoming a pressing commercial issue, because companies—especially SMEs –need to ensure that knowledge and expertise sharing flows in a timely and accurate manner (Gupta et al., 2000; Hau et al., 2013; Nonaka & Von Krogh, 2009; Pipek et al., 2012). As one respondent above suggested, currently knowledge is usually concentrated in the hands of a very few actors. When these actors are for any reason no longer available, say due to ill-health or new job opportunities, a company can be faced with serious difficulties.

So, our aim in this paper is to move a step towards reframing the problem of knowledge and expertise sharing in CSCW. We argue that conventional knowledge transfer models do not suffice to handle what we have observed in industrial contexts, since they do not tackle the embodied work that we have shown to be decisive for effective knowledge and expertise sharing. Our findings here suggest that much knowledge and expertise can be shared by simply observing in adequate detail the embodied action in which practices are embedded.

Much of the data we presented concerns haptics. The work witnessably involves the use of hands, the measurement of pressure and speed, the identification of angles, and so on. The potential for CPS here is evident, because sensors can provide precise haptic information. It is therefore likely that sometime soon embedded sensors will capture certain aspects of practices that are typically described as involving 'know how'. Our argument is that we should take advantage of the affordances of such technology and utilise it as part of a knowledge sharing system. This may well also support workers in dealing with other challenges, such as locating and/or identifying the right parts needed for a set-up process, avoiding time being lost in using incorrect parts that at first sight seem to be correct.

As discussed in our findings, there are parts that differ from each other only very slightly due to the angle or due to millimetre differences. In the case of bending, 1 or 2 mm can make an enormous difference. The discrepancy between a lack of accuracy in the available information and the need for very accurate settings can potentially be addressed by CPSs which, integrate the sensor technology we outline above with (possibly) augmented reality information. This can potentially provide additional information for the man-machine control loop (Gellersen, 2005). In so doing, the decision criteria, on the basis of which the setter defines the correction parameters, is widened. Specifically, this means that non-visible physical processes are made visible to the machine operator through the use of sensors (Chong et al., 2003). This might ensure that the trial-and-error phases are shortened and that the success of the set-up is achieved more quickly.

At this point, we argue that sensors that are integrated into a CPS can play an important role in the knowledge sharing process. Sensors can provide targeted information about discrepancies. A simple example is a temperature sensor. This provides accurate temperature readings at one point in the process. By combining this information with the results of the bending process and establishing correlations, the installer can use this information to set the parameters more quickly in future bends. As part of a CPS, however, sensors are only part of the story. Only the visualization and contextualization of the sensory information represents a relevant added value for the machine operator. To be useable the measured values also need to be processed in an easily understandable manner and displayed in the correct context. Due to the dynamic nature of the setting, real-time systems are advantageous since data can then be accessed without significant delay. Thus, the machine setter is able to allocate the measured values to the part produced, based on the knowledge and expertise shared through the values collected through the sensor technology embedded in the CPS in use. Note in particular how this is embedded within the ongoing accomplishment of the work, rather than demanding overt explication. This addresses both problems of capture and the accountability of person-to-person sharing.

To conclude, the use of AR provides another layer to enrich this knowledge and expertise sharing. Through the use of AR systems, workers can potentially see visual digital representations coming, for instance, from a simulation at the same time they observe the physical machine and parts. This, we argue, has considerable potential to improve the accuracy and timeliness of information and expertise sharing.

4.8 Conclusion and Future Work

Past and current CSCW research, visible in the thorough literature review provided by Ackerman et al. (2013), both show that expertise sharing remains a relevant and timely topic but also that the character of systems to support it will depend on the domain in which they are inserted. CSCW, as yet, has spent relatively little time in the investigation of manufacturing contexts in which such systems might play a significant role and here we have provided some initial results which, we hope, might prompt future research.

We argue that we have whole new ways of recording movement, the haptic, and other types of embodied action, and aligning it with propositional content via AR and sensor technology. This is to-date largely underexplored to our knowledge and has huge implications for the way we think about sharing expertise and knowledge. A first attempt towards a discussion such as the one that we presented here has been made by Ludwig et al. (2017) but the empirical data we provide here adds force to the argument.

With this paper, we take a step forward in reframing the problem of knowledge and expertise sharing. We argue that we are about to enter a third generation of research in the matter and as future work, we set out to pursue this strand of research, especially in regard to the process herein explored, i.e. industrial set-up. Following the premises of the DCS framework (Wulf et al., 2015), we have already started with the design phase of our initiative and carried out a couple of design workshops with our participants. In this phase, we are focusing on the design and implementation of a CPS for knowledge and expertise sharing in industrial set-up, following principles of participatory design and evolutionary development. As soon as we have a stable functional prototype, we will engage in a technology appropriation study. In so doing, we hope to be able to uncover further challenges concerning this third generation of CSCW research on knowledge and expertise sharing.

In particular, we call for research on the challenges associated with the use of CPS. Despite the great potential of these systems, the literature has already identified challenges concerning its use, which go from difficulties in handling new interaction principles to ergonomic problems due to the use of certain devices, e.g. AR glasses (Lee, 2008; Monostori, 2014; Paelke & Röcker, 2015).

It is worth pointing out that, although our findings provide a detailed account of the practices of machine operators carrying out set-up processes, our findings are not immediately generalisable. However, we are confident that our findings about the need to balance information about 'know how' with other forms of knowledge and record it via new technological affordances holds promise on a more general level. e.g. in the context of set-up of stamping machines.

Finally, the use of AR- and sensor-based CPS for industrial set-up is only one of the possible areas of application of such systems. It would be really useful to explore these issues in other contexts, so as to better understand the potentials and limitations of such technologies for knowledge and expertise sharing across a range of domains.

5 Cyber-Physical Systems for Knowledge and Expertise Sharing in Manufacturing Contexts: Towards a Model Enabling Design (JCSCW 2019)

Cyber-physical systems (CPS) are currently attracting a great deal of attention as a part of the discourse surrounding the fourth industrial revolution. Thus far, the chief focus has been upon complex architectures for supply chain-wide data exchange between intelligent machines. Here, however, we take a very different tack by examining the support CPS may offer for the exchange and acquisition of knowledge-intensive human practices in the context of manual set-up processes on modern production machines. The paper offers contributions to both CSCW and the ongoing development of CPS in three ways. Firstly, it presents a detailed empirical study of knowledge and expertise sharing practices in a production environment. Secondly, the results of this study are used as the basis of an overarching model that was developed with the express purpose of facilitating design. Finally, CPS-based technical possibilities are matched to the practice-relevant knowledge and expertise sharing requirements captured within the model.

5.1 Introduction

Knowledge and expertise sharing (KES), also known as knowledge management and transfer in communities such as Information Systems (IS) and Organisational Management (OM), has attracted significant research interest for a number of years. Much of this can be associated with the fact that it has been viewed as a key aspect of organisational competitiveness in the literature by allowing organisations to maintain knowledge assets relating to their products and processes. It can also foster the generation of new knowledge and expertise and improve processes and responsiveness to market demand (Bolisani & Scarso, 1999; Nonaka & Takeuchi, 1995; Randall et al., 1996; Spender, 1996b). Ackerman et al.'s (2013) overview of KES-related literature in CSCW, notes the existence of two distinct

S. Hoffmann, *Knowledge Sharing on the Shopfloor*,
https://doi.org/10.1007/978-3-658-50133-4_5

generations of research: one focused on a repository model for the sharing of knowledge and expertise; the other focused on finding knowledgeable actors and investigating the social interactions with and amongst them. More recently, it has been argued that we are entering a third generation of KES research with the rise of cyber-physical systems (CPS). These, it is suggested, offer new affordances for capturing and displaying knowledge, especially with regard to non-propositional knowledge, which is usually embedded and embodied in action (see Chap. 4).

CPS refers to networked and engineered systems that bring the cybernetic and physical worlds together, with elements of the cybernetic world enabling the monitoring, control and coordination of elements of the physical world, whilst the latter provides the conditions and concrete systems that inform and shape the behaviour of the former (Lee & Seshia, 2015; Rajkumar et al., 2010). They evolved from the concept of embedded systems and are closely related to concepts such as the Internet of Things (IoT), sensor networks and multi-agent systems (Monostori et al., 2016). It has been argued that these systems will result in significant changes across various aspects of work and life by enabling new ways of doing and experiencing daily activities (National Institute of Standards and Technology, 2013). This sanguine vision overlooks some of the challenges associated with this technology. At a technical level this includes things like integrating heterogeneous hardware technology, increased system complexity and the extent to which sensors may or may not be available (Leit et al., 2016; Pacaux-Lemoine et al., 2018; Wang et al., 2015). At a social level there are also potential limitations in how CPS might operate with regard to privacy and at an organisational level there are clearly issues regarding investment. All of this underscores the need to deploy and test such technologies in natural settings to further explore their viability. Despite these prospective issues, however, we see definite potential for this technology to support KES in certain kinds of settings, notably manufacturing. This potential revolves around the possibility of collecting fine grained information about particular work practices and visualising associated activities from a first-person perspective (see Chap. 4)[1].

This paper arises from an ongoing programme of research that is investigating the potential of CPS to support work in manufacturing contexts[2]. All of the work

[1] Clearly, despite the grand vision, CPS will not be able to solve all of the problems with KES. Some of the more organisationally-founded issues will remain beyond its remit. Most importantly, CPS technology is relatively new and has a long way to go before achieving its full potential. In our view, this does not preclude the value of what it may have to offer from a research perspective (see Chap. 4).

[2] In manufacturing contexts CPS have been termed cyber-physical production systems (CPPS) (Monostori et al., 2016). CPPS are just CPS to support manufacturing processes. For

in the paper is empirically-grounded in a Design Case Study (DCS) (Wulf et al., 2015) that has involved extensive ethnographic studies in five Small and Medium Enterprises (SMEs) located in two different European countries. Despite advances in KES research, the literature is surprisingly mute about the specific interactions that take place during the process, especially in an industrial context. Yet, with insights of this order in-hand, it is likely that CPS could offer much in support of it. Additionally, the challenges associated with integrating CPS into processes is currently underexplored in CSCW research.

The contribution offered here is threefold. First and foremost, we provide a nuanced account of KES practices in manufacturing contexts and the social interactions stemming from them. Detailed ethnographic studies of manufacturing, especially of work on the factory floor, are few and far between. As will be discussed in due course, existing studies of KES in manufacturing environments are high-level and provide no insight as to how knowledge is practically shared between workers on the ground. We therefore look at the social relationships between the parties involved in KES and the way people actually interact when sharing knowledge and expertise. In this regard it should also be noted that industrial contexts in general, and manufacturing in particular, are underexplored domains in CSCW. Thus, the research reported here also adds important detail to this limited corpus.

A second contribution of the paper is an examination of how the findings from the above might present design opportunities for CPS and how, in particular, CPS might facilitate KES-related interactions. It will be found that higher-level abstractions of KES in manufacturing settings have already identified some of the issues we examine here, for instance the heavy dependence upon embodied knowledge, typically casts as 'tacit knowledge' in such studies, for certain processes such as set-up. However, these studies fail to delve into the actual ways in which embodied knowledge is articulated and made manifest in practice. Therefore they miss the opportunities for trading upon its manifest properties by pulling upon new technological possibilities, such as CPS. We aim to rectify this omission by considering how AR- and sensor-based CPS might bring about important methodological changes in how KES is accomplished (see Chap. 4). The embodied properties of KES in manufacturing settings are, whilst not wholly unique to factory floors, characteristic of specialized manual work, making it disproportionately important to understand just how it is accomplished through the close observation of the actual work. The work presented here looks specifically

---

simplicity's sake we will use CPS throughout the paper, even when talking about manufacturing contexts.

at the implications of our ethnographic findings for how CPS might be designed to support KES in industrial contexts and will detail the design of a CPS that has been specifically developed for a manufacturing scenario. A key feature of CPS is its capacity to pull upon a range of technological affordances to capture the manifest properties of embodied knowledge and to then be able to convey that in situ in ways that were simply not feasible before. By examining the practical ways in which such a system might be constructed, we are able to further reflect upon the advantages and challenges that might stem from integrating CPS into these kinds of processes.

Thirdly, provide a model that is dedicated to facilitating the articulation of the preceding two concerns. This involves abstraction of both the higher-level elements in the described research and potentially related design activities. The model has been developed with the express purpose of facilitating the design of new and innovative systems for KES and, as such, it offers the possibility of developing insights that may extend the scope of integrating CPS into KES-related research in ways that are fully in keeping with the notion of a third generation of KES research mentioned above. The key benefit that accrues to models of this kind is that they provide an effective and flexible resource for moving between specific ethnographically-founded findings and the practical business of designing and developing systems that reflect the importance of the ethnographic work.

5.2 Related Work

Our consideration of prior research relevant to this paper will begin by looking at the most relevant work in CSCW regarding KES, with a particular emphasis upon two distinct bodies of work, those adopting some form of repository model and those using a more interactionally-focused model. We will briefly examine the paucity of work focused upon how KES is actually accomplished in practice in manufacturing settings, something this paper seeks, in part, to redress. This will be followed by a consideration of relevant studies of KES in other domains. We will then move on to discuss the literature regarding CPS and its potential to inspire a third generation of KES-related work in CSCW. The section concludes with a consideration of how modelling has been used as an aid to design in this kind of context.

5.2.1 CSCW Studies of Knowledge and Expertise Sharing

In CSCW, KES has been a recurrent topic of interest since the late 1980s, when terms like organisational memory, organisational remembering, collective intelligence and expertise location became commonplace (Ackerman et al., 2013; Randall et al., 1996). From the outset a focus was placed on developing systems to support the creation of an 'organisational memory' that could be accessed by whoever needed it and whenever they needed it (Ackerman, 1998). Thus, the first generation of KES research in CSCW focused on a repository model, whose limitations have soon become apparent. Amongst other things, issues regarding the dichotomy between experts and users, problems with the contextualisation of knowledge stored in the system and the authoring burden stemming from the use of such systems emerged (Ackerman & McDonald, 1996). Furthermore, Randall et al. (1996) pointed out that the metaphor of an organisational memory failed to capture the socially situated nuances of processes associated with KES. Hence, attention was drawn to the fact that, whilst organisational memory implies a particular and static conceptualisation of 'skills', 'knowledge' and 'memory', understanding the social context where the process of 'remembering' (as in accessing information and making use of it in some way) actually happens is key to grasping what is going on when KES-related activities are being undertaken. This change in focus originated a second generation of KES research in CSCW, which focused mainly on how to support interactions related to the sharing of knowledge and expertise and the vehicles whereby knowledge is shared amongst knowledgeable actors (Ackerman et al., 2013).

In response to the growing interest in tacit knowledge in the OM (Organisational Management) and IS (Information Systems) literatures, a debate has also developed in the CSCW literature over a number of years about the distinctiveness of the practical skills and knowledge people are drawing upon when undertaking certain kinds of tasks (Ackerman et al., 2013; Schmidt, 2012). The acquisition and realisation of practical tasks that are resistant to any propositional formulation of the knowledge exercised in their accomplishment has been described in the literature as 'know-how' (Ryle, 1945, 2009). Meanwhile, an understanding of tasks that, contrariwise, can be discussed in a propositional form, often in connection with 'know-why' (Lapre et al., 2000), has been described as 'know-that' (Ryle, 1945, 2009). This debate, is of relevance for our own interest in production processes because it impacts on the conveyance of how a range of production activities should be performed and how they might be represented and/or shared with others who might want to engage in similar activities.

Of particular moment for CSCW-related design is how knowledge sharing proceeds in environments where technical tools play an increasingly important role, with a concomitant concern that knowledge and expertise captured from experts should be used and reused as many times as required, by novice or less experienced workers. This underscores the urgent need for high quality studies that examine exactly how the sharing of both propositional and non-propositional knowledge is accomplished in real-world environments. Propositional knowledge represents declarative knowledge, which describes easily understandable facts and is therefore quickly accessible within the context of a knowledge transfer process. Non-propositional knowledge is embodied and often embedded in action, only becoming apparent through particular features of its embodiment or the actions that put it in practice (see Chap. 4).

Numerous studies from a variety of domains have contributed to the field of KES. The work by Randall et al. (1996) regarding organisational is a notable study in this regard, where observations of the critical role played by one particular individual in articulating 'local knowledge' were heavily drawn upon. Another key piece of work in this regard was undertaken at Xerox PARC regarding the sharing of information between service technicians (Yamauchi et al., 2003), itself premised upon an earlier study by Orr (1996). Here, the focus was upon how service technicians went about providing one another with the details relevant to handling exceptional cases when engaged in printer repair. Some work also examines how knowledge sharing is undertaken within and between different scientific communities (Velden, 2013), where much is made of the impact of personal interests when engaged in knowledge sharing activities. Beyond this, there is also work that looks at how documents get shared and re-used as a feature of knowledge work (Harper et al., 2001).

Nevertheless, there are remarkably few CSCW studies that have actively set out to observe just how knowledge sharing gets done in practice in real-world manufacture. There is a smattering of work upon industrial settings and production environments in CSCW, e.g. Button and Sharrock's observations of work in a printshop (1997), Button and Harper's study of a foam manufacturer (1995), Carstensen et al.'s study of production planning in manufacturing (1999), Clarke et al.'s study of a steel mill (2003) and Robinson et al.'s study of a paper mill (2000). None of these, however, focus upon knowledge sharing. This further underscores our argument that the capture, description and analysis of KES in production environments is an important topic of interest for CSCW research, adding an important dimension to the ongoing interest in KES (Ackerman et al., 2013).

Our earlier contribution (see Chap. 4) and this new material together move towards redressing this gap in the literature by helping to develop a corpus of empirically-focused work relating to KES.

5.2.2 Other Relevant Studies of Knowledge and Expertise Sharing

A number of fields have taken an interest in KES in organisational contexts in recent decades. Whilst researchers in CSCW have sought to understand and support the process of sharing, researchers in fields such as OM and IS have focused on processes of knowledge creation and management. Much of the work here draws upon Polanyi's concept of *tacit knowledge*[3], spearheaded since the late 1980s by Nonaka et al. (Nonaka, 1988; Nonaka et al., 2000; Nonaka & Takeuchi, 1995; Nonaka & Von Krogh, 2009). In these studies, organisations are seen as both information-processing machines and as entities that, through the action and interaction of their components, create knowledge (Nonaka et al., 2000). Understanding the essence of this process is argued to be key for successful knowledge management and transfer. To tackle this, Nonaka et al. set out to develop a knowledge-based theory of the firm (ibid.).

The SECI model, as it was termed by the authors, was a spiral-based model that covered four processes of 'knowledge conversion'—*socialisation, externalisation, combination and internalisation* (SECI) (Nonaka et al., 2000). *Socialisation* is the process by which knowledge sharing is negotiated between individuals, usually through direct social interaction, observation and imitation. *Externalisation* describes the process whereby actors attempt to transform implicit knowledge into explicit knowledge through (linguistic) articulation. At the point of combination, explicit knowledge elements are exchanged, linked and enriched at different levels. Here explicit knowledge, often articulated through documents or electronic communication, is exchanged between individuals. During *internalisation*, knowledge seekers attempt to internalise explicit knowledge (Nonaka et al., 2000). This is argued to be a difficult but important aspect of creating new knowledge, only arising when existing (experiential) knowledge has been

[3] There has been extensive debate regarding 'tacit knowledge' in CSCW—see Schmidt (2012) for a thorough discussion of the matter. We will not enter into this discussion here because we are simply concerned with acknowledging research that has been performed in related areas. In Chap. 4, we argue, in line Schmidt's position on KES, that much of what has been deemed tacit is not tacit at all, just embedded in embodied action and visible in other ways apart from propositions.

internalised through processes of socialisation, externalisation and combination (Döring, 2016). Internalisation includes methods such as "learning-by-doing", simulations or experiments (Hoffmann, 2009).

Another relevant aspect of the work in OM and IS is the attention given to the role played in knowledge management when ordinary technologies are used for daily communication, such as E-Mail and video conferencing systems (Bolisani & Scarso, 1999; Goh, 2002). Whilst some of these studies might be regarded as overly optimistic (e.g. Roberts' (2000) critique of Bolisani and Scarso's work (Bolisani & Scarso, 1999)), much of this research draws attention to the limitations of computer technology for achieving effective knowledge management in companies, even where it is acknowledged to be playing a vital role (Watson & Hewett, 2006). Thus, Roberts (2000) critically reviewed the potential of then-available ICTs for KES, especially e-mail, voicemail, teleconferencing, video-conferencing, desktop-based systems, computer-aided design (CAD), computer-aided manufacturing (CAM), information databases and groupware. Roberts' assessment was based on findings from studies of knowledge management and transfer in Research and Development (R&D) activities in multinational enterprises (ibid.). The principal discussion focuses once again upon a perceived dichotomy between 'tacit' and explicit, or codified knowledge. The author concludes that, even if ICTs might facilitate and speed up the collection, storage and retrieval of knowledge in/from so-called knowledge bases, they would still struggle to convey 'tacit' knowledge. In particular, it is argued that such systems would not be able to translate know-how into show-how. Our own studies similarly suggest that conventional computer technologies would not meet the needs of sharing the kind of knowledge encountered in embodied action (see Chap. 4).

Studies of coding communities and their development from mailing lists to web-based Q&A help pages stand apart from this discussion. Here, it has been found that Q&A websites are an effective means of knowledge exchange. Based on questions and answers, knowledge exchange is initiated and incentives created through gamification approaches (Treude et al., 2011; Vasilescu et al., 2013). A positive aspect of this is that the knowledge transfer process accelerates development by broadening the knowledge of knowledge seekers. However, participation in the knowledge community represents an interruption of actual coding activities, which can lead to delays. Studies of stack overflow have shown that gamification approaches used on the platform work to encourage users to participate and be active in the community, leading to an increase in programming productivity (Vasilescu et al., 2014).

Studies in medical settings deal with comparable aspects of expertise sharing in a completely different domain. Normark and Randall (2005) report on the pooling of expertise in an Emergency Call Center that is the organized work of several people, often at different locations or mobile, who seek and offer relevant knowledge in an appropriate form to meet time-critical conditions. Here, systems for knowledge exchange are useful if they support practices that can adapt to organisational need and be accompanied by two key organizational and cultural changes: (1) the development of a culture in which knowledge is organically built and embedded; (2) the development of a strategic role for experts, e.g. shift leaders, who impart knowledge directly or indirectly. Bossen et al. (2013) analysed an IT system that supported home care workers and relatives in cooperative home care work through information exchange and the coordination of tasks. Home care is based on a complex network of activities with a characteristically reciprocal division of labour between care workers and relatives. The IT system helped to provide participants with 'peace of mind' that allowed them to realize their individual engagement. The authors emphasized the efficiency of the IT system and noted its high acceptance by all participants.

In the context of Web 2.0, Morris et al. (2010) have explored the use of social network status messages to ask questions. Their results show that Q&A sites may be more attractive to users who have subjective questions on highly personal topics, since they afford greater anonymity. Search engines, by contrast, provide more objective answers to a variety of personal and impersonal concerns. People tend to trust the opinions of people they know rather than strangers, motivating them to turn to social networks. There is a need here for rapid feedback and a broad spectrum of high-quality answers. In relation to this, it has been suggested that personalized search algorithms or applications may increase user-friendliness and confidence in the system. The use of social media also creates the possibility of company-wide knowledge exchange in work settings by moving away from intermittent, centralized knowledge management processes to continuous online knowledge-related conversations. Little is known about how individuals reconcile their work with knowledge-transfer during limited working hours. This suggests a need to rethink traditional knowledge transfer theories and to consider other approaches to complex knowledge work. Research here has looked at promoting the integration of technologies with individual options for action that bring users and technological artefacts closer together by means of a more "socially acceptable approach" (Majchrzak et al., 2013).

Of relevance here is Yang et al.'s (2008) analysis of a Chinese witkey system that connected users via a knowledge sharing platform and allowed a monetary reward for answers. The aim was to explore underlying properties of users and jobs/tasks to get insights about participation structures. Their results revealed that monetary rewards were not a significant incentive. By contrast, people's perceived expertise significantly affected participation (ibid.). In another Q&A related study, Adamic et al. (2008) found clusters according to the length and overlap of requests and replies. While many users contribute broadly across a variety of themes, technical users tend to answer more selectively. Thus, an indicator of expertise is often the number of answers marked as helpful.

The above studies are clearly relevant to the work we present here, especially our interest in developing a model of KES for CPS design. However, the online and distributed character of the studied phenomena and their focus upon articulating largely propositional knowledge puts them apart from our own interest in how largely non-propositional and embodied knowledge might be effectively conveyed.

5.2.3 Cyber-Physical Systems

We have already noted above our view that it is unlikely that conventional computer technologies will be able to effectively support the sharing of embodied knowledge. Embodied knowledge refers to knowledge that is primarily articulated through embodied action rather than through verbal propositions, though it may be accompanied by verbal propositions as well. So, as we shall be seeing, operating a machine in such a way as to bend a tube to precisely the right degree makes manifest through a series of body movements a particular kind of knowledge and expertise. Previous studies have already indicated that advances in the area of CPS, especially regarding the integration of AR for the visualisation of particular elements, will bring new opportunities for sharing this kind of knowledge. Indeed, as noted above, some have argued that the use of CPS will mark the beginning of a third generation of KES by overstepping existing assumptions about how knowledge might be gathered and visualised (see Chap. 4). Thus, CPS may constitute an opportunity to re-vitalise the design of innovative technologies for KES. In this paper we will concentrate on developments in the area of CPPS because these are most relevant to our interest in supporting the work in production environments.

The broader potential of digital technologies for KES has been widely discussed and exploited within both CSCW and OM (Ackerman et al., 2013; Bolisani & Scarso, 1999; Nonaka & Von Krogh, 2009; Roberts, 2000). In the case of CPPS, however, the burgeoning use of sensor technology offers new opportunities for capturing information about the sequential and haptic characteristics of activities—i.e. the ways in which they are embodied. In manufacturing contexts, this kind of technology can be used to capture things like the delicacy of certain alignments that may turn upon the addition or deduction of miniscule refinements in a machine's configuration (see Chap. 4). AR technology has also demonstrated great potential for supporting the transmission of embodied action by facilitating communication between novices and experienced parties involved in various forms of cooperation (Gauglitz et al., 2012, 2014a). AR has also been found to support unrestricted and context-specific access to information, assisting in the exploration of learning material from different perspectives and enabling teaching where no real-world first-hand experience is possible (Klopfer et al., 2005; Yuen et al., 2011).

At present, CPPS are only just beginning to be exploited. This makes it even more important that appropriate design considerations are taken on-board before fateful decisions are made. For the larger part the existing literature on CPPS is speculative in nature, looking forward to its potential rather than reporting on how it has actually been brought to bear in any specific way (Huang et al., 2017; Jeschke et al., 2017; Schlechtendahl et al., 2015; Zhou et al., 2015). This literature is still largely centred upon industry-related publications. Outside of Chap. 4, it has not yet been seriously touched upon in CSCW. Considerations of how to explicitly build knowledge sharing support into CPPS are still pretty well absent from the literature.

5.2.4 Modelling for Design

One of the objectives of this paper is to present a model that can help to facilitate the building of CPS for KES, including knowledge articulated through embodied action. The modelling literature as a whole is vast and we could not hope to provide an overview in the space available. However, modelling is not a topic that has received the same amount of attention in CSCW. A lot of the literature in this regard is relatively early, as designers sought to get to grips with the implications of the domain. A great deal of this was focused on the modelling of applications (Benford et al., 1993), with modelling related to groupware playing a dominant role (Ellis & Wainer, 1994). There was also a lot of interest in coming up with models of communication at the time (De Mickelis & Grasso, 1994; Simon et al., 1994). More recent work has become more broad-brush, examining the scope for developing generic models to support collaboration and coordination (Fu, 2008; Lee & Paine, 2015; Mainwaring & Lee, 2017). Some of the early work also took as its topic the support of knowledge sharing in cooperative work (Divitini et al., 1993).

Here we are particularly interested in developing a model that can serve as a bridge between our ethnographic findings and the design of CPS. There have been numerous attempts to develop formal artefacts and models that might act as a conduit between ethnography and design over the years. Some approaches have turned this into a near-processual requirement (Holtzblatt & Jones, 2019), but others have also put the role of the model centre-stage as well (see, for instance, Salvador and Mateas's (1997) notions regarding 'design ethnography', where models are derived from ethnographic data to explicitly drive design, an idea that was also subsequently taken on board by Wakkary et al. (2007)). Others have also seen a place for graphical representations as a means of articulating ethnographic work in ways that designers might find it easier to engage with, but have taken issue with the more formalistic notions mentioned above, insisting on the role of the representation as a situated aid to communication, rather than as free-standing artefact in its own right (Diggins & Tolmie, 2003). This has informed other work in CSCW, some of it inspired by Alexander's work on developing a generative grammar for architecture through his notion of a 'pattern language' (Alexander, 1977). In each case, this has involved deriving key features pointing to recurrent social phenomena that might serve as an inspiration for design (Crabtree & Rodden, 2002; Denef & Keyson, 2012; Martin et al., 2002).

There have also been a number of approaches devoted to modelling KES. The best known of these is Nonaka and Takeuchi's (1995) spiral model of knowledge creation, already discussed Peinl's Model of Knowledge Sharing (Peinl, 2006), that also deal with intra-organisational knowledge sharing. Raelin (1997) shifts the emphasis in KES towards learning. He lists learning models embedded in the working environment and distinguishes between two points of view: the individual; and the collective. For both of these perspectives, the existing knowledge base is subdivided into explicit and implicit parts. The learning process is then divided into theoretical and practical aspects. Thus, you may have an individual engaging in the theoretical learning of explicit knowledge. Another may seek the theoretical acquisition of tacit knowledge. Or a collective may adopt a practical approach to the acquisition of explicit knowledge (e.g. 'hands-on' instructed classes), and so on.

Another learning-related model strongly associated with the modelling and design of knowledge systems (Paas et al., 2003; Sweller, 1994) is Cognitive Load Theory (CLT) (Sweller, 1994). This is mainly concerned with learning related to complex tasks. CLT assumes learners have a limited working memory (short-term memory for the recording of information necessary for the execution of a task at hand) when addressing new problems. The construction of new schemes and automation of existing schemes is modelled as a way of allowing an actor to free-up working memory. CLT suggests working memory has three different types of load. The amount of required information and the interaction with it imposed upon the learner is called intrinsic load. Extraneous load is imposed by information that does not foster the derivation of new schemes or that impedes automation. Germane load relates to the building of new or automated schemes. A key aspect of this approach that informs our own model is how different kinds of load may carry different kinds of implications for how learning is realised and thus how the sharing of knowledge might proceed. In particular, we will be relating this to how CPS might best support the capture and visualization of non-propositional knowledge, as discussed in Chap. 4.

5.3 Methodology

The findings leading to the KES model presented in Sect. 5.4 come from an in-depth study that explored the opportunities to design and develop CPS to support machine operators engaged in a time-critical and economically important manufacturing process: machine set-up. Machine set-up refers to the customisation of machines by adding or removing certain components so as to produce different artefacts. This is a key process for manufacturers because it enables the manufacturing of different products by a single machine, at a time of increasing demand for product variability (see Chap. 4). Given the significance of the process, a lot of research has been devoted to its optimization, leading to widely-accepted procedures such as the SMED approach (Shingo, 1985).

To understand how CPS might support user needs, we adopted a DCS approach (Wulf et al., 2015). DCS is a user-centred framework for designing and developing socio-technical systems that investigates how such systems might be appropriated in real scenarios of use. The framework is organised in three different phases: (1) a *Pre-study* seeking to understand the user context and practices and the needs stemming from them; (2) a *Design* phase where new solutions are designed on the basis of the identified needs; (3) an *Appropriation* phase where the design solutions are deployed in a real-world scenario and the outcomes of the deployment closely observed to assess its impact and any resulting social change.

In accordance with the premises of this approach, the investigation began with an extensive ethnographic study, carried out over 10 months (June 2016—April 2017) in five companies based in two different European countries. In this phase, several data collection instruments were used to achieve the necessary richness of understanding regarding the design opportunities we were looking for. In summary, we: (1) carried out a number of observations of several machine operators setting up different machines in different situations, so as to properly understand the process we aimed to support; (2) performed some *shadowing sessions*[4], to understand other issues that could potentially impact upon the machine set-up—e.g. management issues; (3) engaged in a series of in-depth interviews to explore issues emerging from the observations and shadowing sessions; and (4) collected a number of documents concerning the companies' work processes, especially the set-up of various machines—e.g. production plans, process descriptions, machine manuals, etc. Altogether, 25 informants participated in the data collection activities. We involved not only machine operators, but other stakeholders, e.g. process owners, foremen, production engineers, etc. A complete list of participants is provided in Table 5.1. We deliberatively involved participants with different profiles for a better representation of the spectrum of user needs concerning the process we were investigating. Observations lasted from 45 to 90 minutes. Shadowing sessions were up to 4 hours or 3 hours minimum. The observations and shadowing sessions were recorded in fieldnotes. Observations of the process were also recorded using a Tobii Pro Glasses[5] eye tracker, so we could capture fine-grained details of the process and analyse the operator's focus of attention. Interviews lasted from 45 to 120 minutes and were transcribed using the intelligent verbatim method, in which everything is transcribed word for word, except for mumbling (e.g. 'ums' and 'ers'), filler phrases (e.g. 'you know) and stumbles (Hickley, 2003). All data, including the eye-tracker videos[6], was subjected to Thematic Analysis, according to the procedure described by Braun and Clarke

[4] *Shadowing* is a special type of observation, where a researcher shadows the participant, following them wherever they go (Czarniawska, 2007). This technique is of particular relevance for the collection of situated data regarding issues surrounding the main focus of the study. In our case, whilst the observations focused on the set-up process, the shadowing focused on other processes that might have an impact on it. By doing this, we sought to achieve a more nuanced understanding of the social and organizational environment where the process we set out to support was unfolding.

[5] https://www.tobiipro.com/product-listing/tobii-pro-glasses-2/, accessed on 20. October 2018.

[6] In the case of the videos, codes were associated with time frames, just as they were associated with sentences or paragraphs in the transcribed data.

(2012) in which portions of the data are coded by using a mix of bottom-up and top-down approaches. The latter uses a list of codes generated before the start of the analysis, which can come from things like literature reviews or interview guides. A bottom-up approach involves using empirical codes that emerge from close engagement with the data. Later on, themes are generated based on relationships between the identified codes. These are used to answer the research questions governing the investigation. The results of the pre-study have already been reported in Chap. 4. Following the pre-study, we engaged in a design phase. Here, scenario-based design was used to support participatory design. Scenarios were prepared based on analysis of the data collected in the pre-study. These scenarios were used to foster discussion in a series of design workshops carried out to define which types of technologies should be included in our designed solution, the functions it should integrate and its appearance.

After the design workshops, we engaged in rapid prototyping. Early prototypes and the technology chosen to feature in the designed CPS went through a Heuristic Evaluation (Nielsen & Molich, 1990). This was conducted by usability specialists working on the project. Here the goal was to eliminate any well-known usability problems. After improvements arising from the Heuristic Evaluation, the prototypes were subjected to a series of usability tests. Some participants in the fieldwork participated in these preliminary tests, as reported in Table 5.1. This kept them involved in the design process. Other participants were also recruited to test the prototypes (see Table 5.2). These were students from the institute leading the design of the solution. They came from either an industrial engineering background, making them familiar with the process we were trying to support, or from business informatics and thus familiar with the types of technologies we were using. They were selected under the auspices of convenient and purposive sampling strategies (Bryman, 2008)—the former entailing the selection of participants that are easy to access, the latter the selection of participants who might serve a particular purpose, in this case commenting on the usability of the system and its potential to support the process in question.

We are currently performing the last iteration of improvements to the system, before we can engage in the third phase of the DCS. Details of the third phase will be reported in a dedicated publication.

Table 5.1 Fieldwork participants

#	I	E	S	D	T	Role	C	E (u/s)	Exp.
P1	X			X		Foreman	A	graduated (u)	> 10 years
P2	X					Designer	A	Master school (s)	> 10 years
P3	X				X	Foreman	A	Master school (s)	> 10 years
P4	X			X		Machine operator	A	graduated (u)	> 10 years
P5	X	X				Machine operator	A	Master school (s)	< 10 years
P6	X	X		X		Machine operator	A	Master school (s)	< 10 years
P7	X	X	X	X	X	Machine operator	B	Apprenticeship (s)	> 10 years
P8	X	X	X	X		Machine operator	B	Apprenticeship (s)	< 10 years
P9	X	X		X		Machine operator	B	Apprenticeship (s)	< 10 years
P10	X	X		X		Machine operator	B	Apprenticeship (u)	< 10 years
P11	X	X	X	X		Foreman	B	Apprenticeship (u)	< 10 years
P12	X			X		Process Owner	B	Apprenticeship (s)	< 10 years
P13	X			X		Foreman	B	Master school (s)	> 10 years
P14	X					Technical salesperson	B	graduated (s)	> 10 years
P15	X					Technical salesperson	B	Master school (s)	> 10 years
P16	X					Construction engineer	B	graduated (s)	> 10 years
P17	X					Technical salesperson	B	graduated (s)	> 10 years
P18	X					Production engineer	B	graduated (s)	> 10 years
P19	X					Quality engineer	B	graduated (s)	< 10 years
P20	X	X	X	X	X	Foreman	C	Apprenticeship (s)	> 10 years

(continued)

Table 5.1 (continued)

#	I	E	S	D	T	Role	C	E (u/s)	Exp.
P21				X	X	Machine operator	C	Apprenticeship (s)	< 10 years
P22	X	X				Machine operator	D	Apprenticeship (s)	< 10 years
P23	X	X				Forman	D	Master school (s)	> 10 years
P24	X	X				Machine operator	E	Apprenticeship (s)	< 10 years
P25	X					Production engineer	E	graduated (s)	> 10 years

I: Interviews; E: Eye-Tracking; S: Shadowing; D: Design Workshops; T: Preliminary Usability Test; C: Company; E(u/s): Education (unspecialized/specialized)

Table 5.2 Additional Test Participants

#	Background
TP1	Industrial engineering
TP2	Business informatics
TP3	Business informatics
TP4	Business informatics
TP5	Industrial engineering
TP6	Industrial engineering
TP7	Industrial engineering

5.4 Grounds of a KES Model Enabling Design

In this section we present the empirical data that has informed the KES model we introduce in this article. The model itself was constructed to support the design and implementation of new and innovative technologies responding to issues identified in the data. As such, it is just a representation of these issues and the design activities carried out so far (see Fig. 5.1). For coherence, the findings are presented according to the elements in the model. We start by discussing organisational issues surrounding KES and impacting upon it. This is followed by findings concerning the role of knowledge actors in such a process. After this, we look at issues relating to the situational context surrounding KES. Then we examine ways of handling the representation and documentation of knowledge.

Finally, we look at empirical data concerning knowledge seekers and associated issues. The data presented in this section underpins the discussion in Sect. 5.4.1. It also advances the state of the art in its own right, by capturing the nuances of KES processes in industrial contexts, which are currently still underexplored.

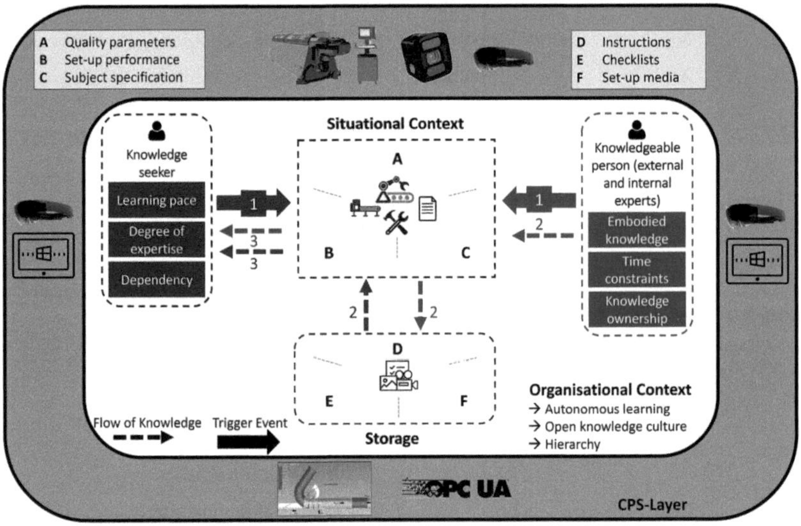

Fig. 5.1 Knowledge and Expertise Sharing Model

5.4.1 Organisational Issues

As can be seen in Fig. 5.1, KES can be impacted by properties of the production system, including aspects of both the working environment and the structure of the organisation. Issues regarding these properties and their effect on KES were a strong feature of the data, as will become clear below.

First and foremost, there are several ways in which workers go about sharing knowledge and expertise with new colleagues. As depicted in the following quote, after showing them how to perform specific tasks and procedures, experienced workers direct novices towards autonomous learning, i.e. a phase where the learner takes over responsibility for the learning process so as to learn key skills independently (Ponton et al., 2005).

 # 1: I tell them: "Here are the tools. If you don't get to grips with where the tools belong, here you have a list." Afterwards I set up the machine together with them about four or five times. Then, one day I go and tell them: I have shown you that several times. Now you make it by yourself. (P7, I)

We observed that novice workers dedicate a great deal of their free time to learning the tools of their trade. At the same time, companies strive to foster an open knowledge culture, so that knowledgeable actors can propagate knowledge acquired within the company. This was reinforced by managers and team leaders, who recurrently said *"it must not happen that someone keeps his knowledge to himself" (# 2, P1, I)*, revealing that hierarchy plays an important role in KES.

Issues of hierarchy not only regulate KES initiatives, but also the appropriation of shared knowledge. Hence, *"[t]here are also differences [in the way knowledge will be appropriated], as [one] pass[es] it on to the employee, who has to prepare afterwards" (# 3, P1, I)*. However, although new workers have some autonomy to shape their learning processes and appropriate shared knowledge, hierarchical interventions can determine how this knowledge is applied. *"They would fetch a colleague and tell him: 'from now on you will make this exactly like I do'" (# 4, P20, D)*

Another issue concerning hierarchy relates to divergent KES expectations. Whilst keeping knowledge to oneself may be seen as a defence mechanism against becoming disposable, managers sometimes nevertheless assume sharing will happen. Thus, it was not uncommon to hear participants in higher hierarchical positions say *"[if] you ask some colleagues [in the company] who have been [there] for a long time, then they will enjoy transferring their knowledge" (# 5, P5, I)*. However, some managers were aware of the difficulties concerning KES. According to them, much of this was to do with ego and was seen as going against the team spirit:

 # 6: Once I had an operator, who kept everything to himself. We had to tear nearly every answer out of him. But, we couldn't try to usurp his throne. This doesn't achieve anything. (P3, I)

In terms of company initiatives for KES, the strategies used include intern workshops and training sessions, where more experienced workers are brought in to share their knowledge and expertise. Another strategy is to contract specialised external training services to provide more formal learning sessions. Interestingly, some participants said that the learning they got from outsiders was *"always better than internal training sessions" (# 7, P1, I)*. This, we think, has to do with several different things. First and foremost, outside training services are paid—often by the company, but occasionally by the workers who are proactively seeking specialisation opportunities—and they are expected and strive to deliver a good service, as this is their business. Therefore, it is sensible to think that there will be investment in people who can really deliver the content, or in other words, whose didactic practices are well-developed[7]. Secondly, their business is about realising effective teaching with cross-company expertise.

5.4.2 The Essential Role of Knowledgeable Persons

The success of KES processes depends on several aspects. First of all, the ability of knowledgeable people to share knowledge (see Fig. 5.1, red items), i.e. the extent to which the didactic practices (Schmidt, 2012) of knowledgeable people are effective, is positively correlated with the extent to which KES will succeed. So, professional, social and communicative competences regarding the introduction of new knowledge, the supervision of how it is applied in practice, and the capacity to intervene when necessary, are basic prerequisites for its success.

Furthermore, the type of knowledge that comes with experience is implicitly bound up with the people themselves, such that knowledge becomes embodied in their practices. This kind of knowledge is difficult to share verbally or by means of propositions, e.g. in the course of set-up processes, parameters have to be defined that determine the movements of the machine axis. This has a direct impact on the quality of the manufactured product.

However, knowledge or know-how about set-up procedures, can often been shared through observation, even if sharing through verbal expressions or propositions might appear difficult (Nonaka et al., 2000). Our findings suggest that, with the help of some "learning-by-doing" practice, this kind of knowledge sharing can be facilitated. Drawing upon knowledgeable people, the values are empirically determined, then internalized by experience during a practical learning phase.

> *# 8: You have to keep in mind that if you start afresh as a machine setter, you will have longer set-up times. However, the longer you work here and the more things you acquire through training, education etc., the shorter the set-up will be. A lot of them are empirical values, which cannot all be passed on or explained simply like that. (P5, I)*

[7] Didactic practices describe the ability of a knowledgeable person to effectively communicate knowledge relating to a specific job by demonstrating it to a knowledge seeker in the course of a knowledge transfer process (Schmidt, 2012).

Automating parts of the recording of information for posterior processing may provide a way of facilitating knowledge sharing among those who are otherwise resistant to engaging in such practices. Several participants, including P3 in the quote below, reported that sometimes people decide not to engage in documentation practices because they want to secure a unique selling point and their position in the company:

> *# 9: There are employees who keep a lot to themselves. They want to use it to secure their job or their position. Maybe it's human, but it doesn't help the company. (P3, I)*

Naturally, automating the recording of information and/or knowledge raises issues of 'ownership' and other ethical issues, which must be taken into consideration and addressed. However, these issues are outside the scope of this paper.

As widely acknowledged in the literature, time is a limited resource in a variety of contexts and production is no exception (Vasilescu et al., 2013). It was not uncommon to hear people say *"those who should transfer knowledge, haven't got enough time to do it" (# 10, P11, I)*. Therefore, there is a certain necessity about the fact that, whilst knowledgeable people may initially have to shape the learning process, knowledge seekers have to gain autonomy in the process as it progresses.

5.4.3 Situated KES

An important aspect of KES is its bidirectional nature. In the literature, Szulanski (2000) summarise the interdependencies of a knowledge sharing process according to four stages: (1) initiation of knowledge; (2) implementation; (3) intensification; and (4) integration of knowledge by the knowledge seeker. In CSCW, the importance of identifying knowledge gaps and knowledge needs and the interaction between the involved actors has been widely acknowledged (Ackerman et al., 2013). Identifying knowledge gaps helps to concretize the goals of knowledge sharing and to shape the sharing activities accordingly. At the end of the knowledge sharing process the knowledge gap also needs to be bridged. As shown in the model, there are three key properties from the empirical data that refer specifically to the situational aspects of KES: quality parameters; set-up performance; and subject specifications.

In a production environment, the quality parameters are a decisive indicator for KES. The execution times for certain activities, in particular manual activities, are also critical. These are compared with target times to assess development and performance potential, but at the same time, they can represent a trigger that initiates knowledge (first phase) sharing (Szulanski, 2000). If one examines the results of the fieldwork presented here, an essential starting point of a transfer process is the realization that temporal differences exist. As one of our fieldwork participants put it, whilst *"one operator needs 30 minutes, the other [might need] 60 minutes" (# 11, P20, I)*. This suggests that the temporal performance of the machine setters can be used as a basis of comparison. If this comparison is expanded and the development of an operator's set-up times and error frequencies is taken into consideration, important conclusions can be drawn about the initiation of knowledge sharing.

> *# 12: If set-up times become shorter, this already shows that they [the operators] are developing a little in this respect. (P20, I)*

If the flow of knowledge is insufficient, workers will lack the necessary knowledge to quickly correct their errors. Therefore, information about the worker's performance can be seen as a trigger for the (first phase) sharing process. It can also be taken as an opportunity to distribute certain tasks, which might contribute to the building of a broader repository of knowledge, as seen in the quote below:

> *# 13: If an instructor discovers a need for knowledge and notices that the operator has difficulties working on certain processes with his existing knowledge, he gives the employee appropriate tasks. By completing these tasks, the trainee can catch up on knowledge and close certain knowledge gaps. (P20, I)*

We were also reminded by participants that, aside from the practice-oriented aspects of knowledge sharing, theoretical foundations in the form of subject-specific and analytical knowledge regarding the manufacturing process are also important. This knowledge is especially important in situation-specific contexts, where it can help with understanding what the implications of a specific situation might be:

> *# 14: ... theoretical and physical knowledge is also perceived as important, so that one knows what changes one has to make. In addition, one can then estimate the consequences and effects of these [changes]. (P11, I)*

5.4.4 Storage Elements

A further organisational strategy for KES is fostering the documentation of work processes and solutions for contingent situations. So, companies still try to build an organisational memory of relevant processes and solutions. Thus, KES may no longer seem solely dependent on the workers' knowledge, as indicated below:

> *# 15: ... some special set-up processes are heavily documented ... this doesn't depend only on the knowledge of the operators, but there it is in black and white, even an operating instruction. (P4, I)*

However, this trend towards documentation in KES practices is not related to making workers disposable. On the contrary, we observed an appreciation for experienced workers and their retention:

> *# 16: There is, I would like to say, no second Mr. [O.] and no replacement for him, and if this happens this way, ... that you even have to search for who situationally knows best in the organization and how to proceed, well ... (P4, I)*

Documentation also has social characteristics in KES. It is a way of recording experienced-based knowledge, so that know-how can be shared (Ackerman et al., 2013; Randall et al., 1996). Participants in our study mentioned recurrent attempts to put this into practice so to keep knowledge available to any interested parties—e.g. novice workers—whilst relieving the demands placed on experienced workers. However, we also saw that the documentation process usually involves a significant investment of time on the part of knowledgeable people. Moreover, written instructions do not replace the communicative and practical approach described above, but rather support it.

This provides another notable opportunity for the technological support of KES, especially with regard to CPS. CPS can automate the collection of some sorts of information that can subsequently be processed and made into consumable knowledge—e.g. by recording via sensor technology the most-used parameters to set up certain machines for the production of specific articles from a particular material, or by using AR technology to facilitate the recording and demonstration of knowledge embodied in action.

Although the repository model has been criticised within the CSCW community, it is worth noting that KES often involves issues of gathering and making knowledge available (Ackerman et al., 2013; Ackerman & Malone, 1990). Our observations also suggest that the reaction to triggering events is often a second phase of knowledge sharing. In this phase, sharing often takes place by means of direct communication. In that case, the second phase passes directly on to the third phase of knowledge acquisition. Alternatively, second phase knowledge transfer may be realized independently of other people through the use of storage technologies. Here, we argue, a step-by-step guide based on checklists can help to reduce the burden on both knowledgeable people and knowledge seekers. This presents a design opportunity for new and innovative technologies, which can benefit the overall process, as suggested by P11.

> *# 17: We could have a simple screen on which the new employee could watch an animation. Step by step, what has to be dismantled and then reattached. That would save time. The [installers] could understand the whole thing more easily with this program. (P11, I)*

Having said this, direct communication with knowledgeable actors was recurrently seen to be a more effective means of correcting errors than written documents, for instance because more could be conveyed through direct communication and it allowed for direct reference to the actual context:

> *# 18: In contrast to documents, direct communication contains more information and refers directly to employee questions. (P7, I)*

However, when working in shifts, direct communication is not always possible. This means that some sort of media support must be provided so that the flow of knowledge can be maintained. Again, this is an opportunity for the design of CPS technologies, which might facilitate the asynchronous distribution of knowledge, as noted below:

> *# 19: Descriptions in the form of photos, texts or video recordings should be available to support novices in their work. With the help of the videos you can see how the experienced employees work. (P4, I).*

Most importantly, these technologies must support the integration of personal experience into the process, so that process-relevant knowledge can be established. Experiences are individual and relate to different situations (Nonaka, 1991; Randall et al., 1996; Spender, 1996a). Therefore they cannot simply be replaced by external assistance. Rather, a coupling of interpreted information, individual knowledge and one's own experiences imbues knowledge with specific relevance and enables it to be integrated.

 # 20: The employee makes these changes and corrections on the basis of his experience gained over many years and many work processes. The employees do not use documents or other tools for this purpose. They can name the causes, make corrections to the machine program and then explain why these changes are made. (P7, I)

5.4.5 Seeking Knowledge and Associated Issues

Our observations suggest that the absorption capacity of those seeking knowledge is fundamental to how far they can handle newly shared knowledge. Put another way, workers must be able to learn new content, combine it with known content, and ultimately understand and apply it in practice (Doetsch, 2016; Hendriks, 1999; Pisano, 1996). In the model (see the blue items in Fig. 5.1) knowledge acquisition and knowledge integration make up what is termed the third phase.

There is an issue, here, regarding the pace of learning. Participants often mention that some people get to grips with something quickly, whilst others need more time: *"Every person is different. You cannot put everyone on the same level" (# 22, P1, I).* However, as time goes by, the learning process gains momentum and things come more naturally, as P7 suggests:

 # 23: I look at some part and, through my routine, I know how it has to be adapted. With new colleagues, it takes them longer, because they don't yet know where to put what. But that is also "learning-by-doing". The more often they do this, the more automatic it will be at some point. (P7, I)

A knowledge seeker's degree of expertise is decisive for the success of knowledge transfer when interacting with a knowledgeable person. For P1, this is because knowledgeable people sometimes overlook mentioning certain details because they take them for granted:

 # 24: There are also people who take some things for granted because they have been with us for 20 years. Then they may forget to mention some little things, which are however important. (P1, I)

The possible limitations of knowledge seekers need to be taken into account. KES for experienced or inexperienced workers can differ significantly, especially in terms of the level of detail that must be provided and the provision of embodied knowledge:

#25: Dealing with trainees should be different from dealing with skilled workers. There is no point in explaining three steps to inexperienced employees at the same time if they have little knowledge about the process. (P20, I)

At the beginning of a knowledge seeker's learning phase, depending on their learning speed, knowledge seekers may be very dependent upon knowledgeable people. This dependency can lead to production bottlenecks if work shifts are not staffed with the right people:

 # 26: Some machine operators cannot react to certain situations. They are completely dependent on us. If they find that the machine has completely misaligned for any reason, they have to wait until someone comes and corrects it. (P6, I)

The relationship between people involved in KES is also influenced by experience (Argote & Ingram, 2000). Knowledgeable people have process- and product-specific experience that was viewed by some participants as a basis of authority with regard to issues concerning the work, as seen in the following:

 # 27: Of all the machine setters, I've been here the longest. As I am also involved in machine maintenance and inspections, I have acquired many things, including tips and tricks. I think I can claim that I know the machines best. (P7, I)

In this article, we have sought to bring all of the nuances and particularities of KES together in one sensible representation that can support the design of new and innovative technologies to support it. We will now discuss certain aspects of the interactions taking place during KES and how they might potentially be mediated by a range of computer technologies that bring together the cybernetic and physical worlds.

5.5 Discussion

Drawing upon both the above material and our previous research regarding KES in manufacturing contexts (see Chap. 4), we have sought to represent our understanding of the particularities of this process and how technologies might support it, so as to render it more tractable for design.

The model combines a very practice-oriented view of the sharing process, based on the empirical data and other material from the literature, in particular existing models for KES (see Sect. 5.2.2). Our reasons for moving towards modelling are entirely pragmatic. First of all, there is a certain inevitability about the fact that some kind of model will be established when designing a relevant technology to support the activities we have observed. Secondly, we recognise the difficulty of getting to grips with the rich ethnographic materials we have gathered for designers. Thus, as per the literature discussed in Sect. 5.2.4, we are seeking to facilitate this by having to-hand an abstraction that captures the key aspects of the ethnographic findings that need to be conveyed, in a format that already posits their relevance to design. Design, of course, can be many things, so the model has been developed in such a way that it can equally well support the design of an actual assistance system, associated applications and services, or simply the creation of an underlying data model.

A key step in developing the model was the identification of relevant stakeholders and an explication of how knowledge sharing relates to their activities across a gamut of social, organisational and environmental issues. Ultimately, only experts can successfully carry out knowledge-intensive processes and technological support can only provide a framework. At best, the framework will allow for consistency in its handling of the process, whilst maintaining the scope for its content to be situationally coherent for specific parties with specific kinds of experience, engaged in specific practices that inevitably evolve over time. We have therefore set about using our ethnographic work and existing approaches to model the core processes, whilst making it possible to select a variety of technologies and define the sharing of a wide range of content. In particular, we have sought to illustrate how CPS might be able to support the modelled process.

5.5.1 The Human as the Start and End Point of the Sharing Process

As illustrated in Fig. 5.1, knowledge sharing takes place between at least two actors—a knowledgeable person and the person learning from her/him (Ackermann et al., 2018; Hoffmann, 2009; Nonaka & Von Krogh, 2009). In existing models, the knowledgeable person is often conceptualized as both a carrier and a source of knowledge who is capable of satisfying/servicing some knowledge seeker's need for knowledge in the pursuit of some particular activity or task (Doetsch, 2016; Nonaka et al., 2000). In relation to this, these models typically recognise a commonsensically obvious fact: sometimes knowledgeable parties notice that someone else needs their help; and sometimes people needing help hunt out other parties who might be able to help them (Ackerman et al., 2013). These have been worked up formally into the following two principles:

- The pull-principle: where a knowledge seeker deliberately seeks out a knowledgeable person based on their existing need for knowledge.
- The push-principle: where a knowledgeable person recognises that they might be able to supply a knowledge seeker with knowledge, because they have noticed a possible need.

Our own model recognizes these two principles by presenting them in phase one, where initiation of the sharing process takes place as a result of some party noticing there is a knowledge gap. This can be the knowledge seeker or the knowledgeable person. Either way, an absence is seen and someone tries to do something about it. The common space where the knowledge seeker and the knowledgeable person come together is at the point of knowledge application, which is, of course, some real situation (e.g. in the course of production) in a real setting (see Sect. 5.4.1).

Both knowledgeable people and knowledge seekers have individual characteristics that can influence the sharing process. So, we saw in our findings that knowledgeable people can be distinguished by the expertise they have to offer and that one way in which the status of a 'knowledgeable person' might be achieved is through seniority. Another way this status might be achieved is by coming onto the site as a 'knowledgeable' external party, for instance, to undertake maintenance or inspect the production facilities. The presence of knowledgeable people both within and beyond the production site is recognised in our model through the distinction between internal and external experts in the entity labelled 'knowledge holder'. Our findings have also emphasised that knowledge seekers can be quite different from one another, most crudely at the level of expertise they already possess, which can be decisive for how knowledge sharing might proceed and with what success (Sweller, 1994).

5.5.2 Interaction Between the Situational and Organizational Context and Data Storage Structures

The findings presented in Sect. 5.4 and its sub-sections show how the situated work of production forms the heart of the sharing context in our model. The work context is of enormous significance for how the communication between a knowledge seeker and a knowledgeable person will unfold. The application of knowledge by both parties involves synchronization with both the specific production context, in all of its detail, and with the boundary conditions of the environment that influence how learning might proceed. There will always be some specific production situation forming the background to knowledge sharing. This could be the machine being worked on, the tools being used, the material being processed, and so on. The situational context also includes information and data regarding the production elements mentioned above and the organizational structure within which the process is embedded. This might involve things like orders, plans, speeds, machine programs or tolerance values. Employees are oriented to these values and they influence both working methods and the results of the work. All of these elements represent the specific problem confronting a production worker in a particular situation, which then needs to be mastered through whatever knowledge is available. The situational context also includes difficulties and frequently encountered problems. For example, setting up a machine for a particular production process can be very complex because the material is very sensitive, thus demanding lots of experience regarding how to set up and operate the machine for the process in hand.

We saw in Sect. 5.4 that a practitioners' view of a specific situation and its surrounding circumstances can also be conceived of as a learning environment when getting to grips with something like a set-up process. Drawing upon Raelin (1997), we can see that the learning environment can be fateful for how something like a set-up process might be achieved. So, one way in which embodied knowledge might be acquired over time is through practicing an activity in a non-consequential setting, what Raelin terms an 'experimental environment'. Knowledge acquired in this way can then be applied in the actual doing of the work, what Raelin calls a 'practice-oriented environment'. However, more needs to be done because, in reality, practicing how to perform a task offline is very different from doing it in a real production environment, where things are changing constantly. Thus, things you've learnt to do in one way may have to be adapted to other circumstances that are only superficially the same but contingently divergent in any number of ways. Worse still, when trying things out in an 'experimental environment', you have the leisure of being able to take pause and reflect upon what you are doing, but this luxury is not typically available in real-world production environments with real-world production schedules (see Sect. 5.4.2).

In Sect. 5.4.2 we also saw that the first step of a learning process is often controlled by direct and personal feedback, on the basis of attempting some particular kind of set-up and with the support of set-up documentation. With the help of the feedback, it is possible to reflect upon the activity, which, according to Banker et al. (1993) and Stone and Stone (1984), leads to what might termed 'know-why' (Banker et al., 1993; Lapre et al., 2000). This can then lead to a phase in the learning process where autonomous learning can largely replace personal feedback. At a certain point, enough is understood for people to have an abstract grasp of what is going on that can be applied across a variety of situations in the production process.

Another thing we noted in the findings is that the frequency of feedback is determined by the amount of direct contact between a knowledgeable person and a knowledge seeker. This need for regular contact is important because, as Wollersheim et al. (2015) have previously pointed out, for complex tasks a lot of feedback is needed to advance learning. This also resonates with views from CLT where it is suggested that immediate feedback helps with the formation of higher-level schemes. In Sect. 5.4 we encountered workers who emphasised the extent to which direct support was appreciated. From the point of view of the model, we can conclude that technologies for KES, be them CPS or any other type of technology, need to support the provision of significant amounts of feedback (Wollersheim et al., 2015). We can also see how CPS may have specific advantages to offer here because, as observed above, the opportunities for feedback and reflection in existing production environments can be limited. By having feedback come from the system itself, fewer resources are tied up. This also offers more than existing forms of offline, leisure-timed based learning, because it can preserve the relationship with specific production contexts.

At the same time, however, we saw workers pointing to the importance of understanding the physical principles underlying the basic mechanisms involved in the set-up process. In other words, there is also a need for knowledge that can transcend specific situations and be applied across a range of different circumstances. This resonates with the view in CLT that there is a need to acquire what it terms 'germane load' that, through the way in which it can conceptualize the application of knowledge to other similar processes, is able to reduce 'cognitive load' (Paas et al., 2003). Over time, this can serve to support the development of expertise (Kalyuga, 2008). Thus, a further design principle that needs to be observed as we move from the findings in the model to the CPS layer, is the need to support the acquisition of location-independent learning.

In general, an open organisational design is required for KES. The aim is to make knowledge freely accessible. However, various considerations can play a part in determining how realistic an objective this actually is. One such consideration is the personal interests of knowledgeable people. Another is the technical limitations on what can be made available in this way. A completely open knowledge environment presupposes that all knowledge is always accessible, with only a minimal need for personal interaction. We encountered divergent views upon this in the findings. In summary, an assistive tool for knowledge sharing needs to be able to indicate the reliability of the knowledge, perhaps with an option for representing degrees of confidence. Furthermore, as it seems unrealistic to expect organisational cultures to become invested with unreserved mutual trust overnight, the assistive tool also needs to provide for the protection of private knowledge against its potential misuse.

Of course, as is recognised in many knowledge sharing models, at a certain point knowledge seekers acquire sufficient expertise for them to become themselves potential knowledgeable persons. To take a parallel example from studies of communities of practices (Lave & Wenger, 1991), knowledge seekers, who at the beginning find themselves at the peripheral boundaries of the community, move to its centre as they become more knowledgeable. Raelin (1997) also points out that an essential part of workplace learning relates to experience. We would add to this that experience is made up of handling a variety of different situational contexts such that key correlations and dependencies can be grasped. This needs to be reflected in how storage elements are designed. Evidently, storage needs to allow for adaptation and flexibility, not just the provision of fixed content.

Organisational structures also play an important part in fostering the conditions under which sharing might be carried out. As we saw in the findings, external learning environments are sometimes preferred because of their independence and the fact that they free participants from local constraints. Thus, participants can escape the ever-present time pressures that have such an impact on people's interactions and knowledge flows. Simply put, in most industrial settings, the sharing of knowledge is in direct competition with the completion of production processes within a restricted amount of time. This limits the number of knowledgeable people who might be available to share their expertise and it restricts the scope for would-be knowledge seekers to fully avail themselves of the knowledge that is ostensibly available within the environment. In view of the kind of knowledge people are trying to acquire for activities such as set-up, leisure time learning does little to offset the problem here. This calls out for the provision of supporting media that can step in when knowledgeable parties are not available. Unfortunately, existing media are not well-suited to fulfilling this role because they lack the capacity to handle feedback requests, thus limiting the formation of 'know-why'. This, again is where CPS has the potential to step in and provide knowledge seekers with more interrogable forms of support.

In relation to the above, it can be seen that data storage structures need to be adaptable to specific situations and organizational contexts such that they can capture situations and imbue them with relevant additional instructional affordances. Instructions need to be embedded in situational contexts so that relevant information can be quickly found and acted upon. At the same time, just what is captured and what is provided in the way of instruction needs handling with care. Sometimes personal preferences may lead to a disregard of the actual situational context. Thus, reference to the actual situation must always rest paramount, both when capturing the data and when using it.

5.5.3 Issues in the Flow of Knowledge

Having introduced the principal elements of our knowledge sharing model, we now move on to elaborating on how those elements are connected to each other. The linking of the elements effectively describes the actual sharing process, as will become clear. From our observations, we identified that in this kind of process, knowledge application and knowledge flow sit side by side. The former concerns the application of one's own knowledge in practice—in our case the production process. The latter refers to the dissemination of knowledge as KES unfolds and travels between knowledgeable people, storage media and knowledge seekers.

Our findings suggest that the push-pull principles described in Sect. 5.4.2 are directly related to the application of knowledge. As seen in Sect. 5.4.5, in order to trigger knowledge sharing, knowledge seekers must proactively ask for information. Moreover, the boundary conditions of the situational context are very important for the organisation and pursuit of KES, as the findings make clear. In our case, the situational context was delimited by the manufacturing context we were exploring, which involved machines, tools, certain kinds of material, etc.

When it comes to knowledge flow, this can happen directly (through personal contact between a knowledgeable person and a knowledge seeker) or indirectly with the help of storage media. In the case of indirect interactions, it is important that embodied knowledge be conveyed (see Chap. 4). This is where CPS comes into its own, by offering opportunities to adapt the communication between knowledge seekers and knowledgeable actors in ways that were previously not possible. Key here is its capacity to facilitate the collection of the necessary knowledge and present it in different formats, e.g. textual representations and static or dynamic graphical representations such as pictures, videos and AR overlays. As suggested in the literature, the possibility of different presentation formats is elementary to the success of sharing, especially in technical settings (Patterson et al., 2014; Wright, 1977).

Documentation can be helpful, but it is often unable to show all of the relevant detail. We also saw in Sect. 5.4 that producing and accessing documentation can be challenging, for a variety of reasons such as time pressure, unwillingness to share, difficulties in finding the relevant information, etc. On the other hand, documentation supports expandability, replicability, standardisation of the learning content and a reduction of certain kinds of effort or, as Sweller (1994) put it, workload. As it provides an abstraction of what is already deemed to be salient for the task in-hand, it can be seen to support what Sweller (op. cit.) calls 'germane workload', which is argued to have a positive effect on limited working memory (Sweller, 1994). It can also be effective in terms of anticipating the degree of expertise of knowledge seekers and their limitations, something else that was noted in the findings.

The core challenge confronting KES that is a recurrent feature of the discussion framed around notions such as tacit knowledge and know-how, is that empirical values concerning set-up parameters, i.e. values that are dependent on the situational context, cannot always be easily conveyed, because they are a product of individual experience. This has led some to argue that, at times, there is no flow of knowledge in the KES process because there are moments where a person is both a knowledge seeker and a conveyor of knowledge. Concretely, there are moments where people 'know the basics', but cannot be said to have acquired mastery until they have 'tried it out' and maybe 'practised it until they've got the hang of it'. We have discussed this already in terms of the potential need for 'experimental environments', something that can be hard to provide in conventional production settings. As we shall discuss further below, this is also potentially rectifiable through the use of CPS.

5.5.4 CPS Layer: Possibilities of Modern Technologies

Earlier in this paper we noted the claimed potential for CPS to bring about significant improvements in the realisation of KES through the use of new technologies. A particular focus here has been upon the use of novel visualizations, such as AR (see Chap. 4). Embedding such visualizations in overall architectures and connecting them up to sensors and machine-learning algorithms is usually discussed in the literature under the auspices of inter-machine communication (Lee et al.,

2015). The results presented in Sect. 5.4, however, suggest that it is human beings themselves and their abilities that are the decisive factor in implementing set-up processes. Nonetheless, if these architecture are set against the background of the Fourth Industrial Revolution and put in conjunction with structures that give primacy to human action, they offer a definite added value, especially for things like the set-up process (Wang, 2010). This goes hand-in-hand with a differentiated approach to the selection of technologies and a smooth integration of operators' interactions with the technology during the actual set-up workflow.

For reasons of clarity, the following Table 5.3 compiles the findings in the series of mentions in the fourth chapter, their source, the implications derived from them and the technical solutions listed in this chapter and refers directly to the figures.

Basic requirements, such as the fact that hands cannot be used continuously during the interaction, are relatively obvious. However, designers also need to focus on ensuring that interaction with the tool will not cause any additional time stress. There is a strong correlation between time and cost in production environments, so processes should not be extended. These requirements can be realized through the interaction possibilities presented by HoloLens technology, with an interaction design that is properly adapted to the set-up process (see for example Table 5.3 No. 7). Figure 5.2[c] shows two of the central principles in our designed solution. One of these is the presence of gesture-based controls, which the HoloLens offers as standard. The other is interactive surfaces. Thus, in our current solution, a menu has been designed that can be interacted with constantly and quickly, without restricting the view of the operator. This is achieved by placing the menu directly in front of the operator on the floor.

Visualizations are embedded in the actual context by merging them with the environment. Ideally, they will also display context-sensitive instructions and information (Zhu et al., 2014). Context-dependent representations can be designed in such a way as to ensure that the instructions—visualized by means of AR glasses—can be found and enacted quickly. All actors will need to be equipped with this technology because the requirements apply equally to all of them. In that case, it needs to be possible for experts to prepare the instructions and make them available independently of any one person, whilst ensuring they remain applicable to specific contexts. The HoloLens is the primary vehicle for this, both storing the set-up steps and displaying the set-up steps through contextualized AR-visualizations. The visualizations are divided into two parts. One part creates and reproduces the exact assembly location and the assembly interaction in the form of an animation that can be displayed on the real machine (see Fig. 5.2 [a]). The operator can use the other part to save the set-up steps as a

Table 5.3 Overview of the empirical findings and the corresponding source as well as the derived design implications with reference to the CPS-Layer

No.	Empirical finding	Source	Implication	CPS-Layer	
1	Autonomy	Ponton et al., 2005; P7	- Autonomous storing and replaying of set-up instructions	HoloLens	Figure 7 [b]
2	Open knowledge culture	Bossen et al., 2013; P1; P3	- Collaborative editing of set-up instructions	Editor	Figure 8
3	Hierarchy	P1; P5; P3	- Involvement of all relevant knowledgeable persons	Editor	Figure 8
4	External experts	P1	- Knowledge integration of external experts	HoloLens	Figure 7 [a,b]
5	Embodied knowledge	Argote and Ingram, 2000; Section 4; P5	- Provide instructions independently of persons and locations - Different forms of visualisation - Sensor based movement tracking - Senosrical monitoring the set-up steps - Record handling results in the form of empirical values	HoloLens, 3D-Camera	Figure 7 [a,b,d] Figure 9
6	Knowledge ownership	Vasilescu et al., 2013; P3	- Provide instructions independently of persons and locations - Offer gamification feeling to support participation	HoloLens	Figure 7 [a,b]
7	Time constrains	Vasilescu et al., 2013; P11	- Adapted interaction for a set-up process - Easy access to context dependend instuctions	HoloLens	Figure 7 [a,c]
8	Quality parameters	P20	- Provide quality parameter feedback to support decision finding - Increase reliability of knowledge	3D-Camera, Virtual machine and product data	Figure 9 Figure 10
9	Set-up performance	P20	- Provide set-up performance feedback as trigger events for expertise sharing process	Editor	Figure 8
10	Subject specification	P11	- Provide virtual set-up process simulation data	Virtual machine and product data	Figure 10
11	Instructions	Ackerman et al., 2013; Randall et al., 1996; P4	- Provide instructions for the set-up processes	HoloLens, Editor	Figure 7 [a,b,c] Figure 8
12	Checklists	P11	- Step by step instructions of set-up	HoloLens	Figure 8 [a,b]
13	Set-up media	P4; P7	- Provide medial support for the set-up process	HoloLens	Figure 7 [b]
14	Learning pace	Doetsch, 2016; Hendriks, 1999; Pisano, 1996; P1; P7	- Instructions have to adapt to the learning speed of the knowledge seeker	HoloLens	Figure 7 [a,b]
15	Degree of expertise	Sweller, 1994; Argote and Ingram, 2000; P1; P20	- Offer different forms of visualization - Knowledge seeker decides on the details	HoloLens	Figure 7 [a,b]
16	Dependency	P6	- Provide instructions independently of persons and locations	HoloLens	Figure 7 [a,b]

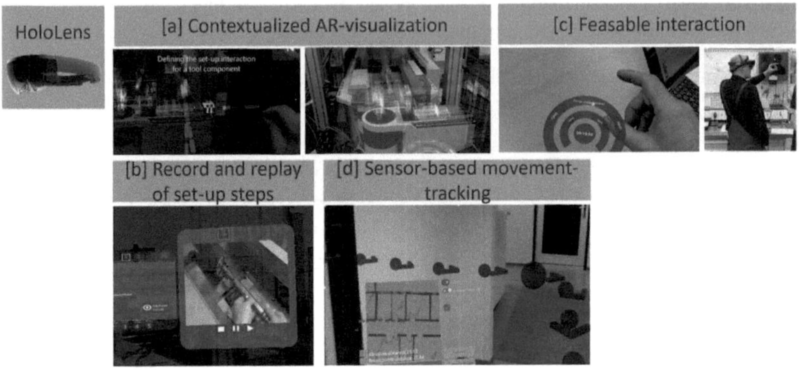

Fig. 5.2 HoloLens for recording and reproducing the set-up steps and as a sensor for recording the logistical processes

video recording so that they can be played back afterwards (see Fig. 5.2 [b]). Thus, the knowledge seeker is able to select autonomously the scope and level of detail of the instructions, either using just the AR instructions as a basis for the execution of the step, or working with all the video instructions. The adaptability of the set-up instructions through different forms of visualization reflects the need to be able to match the characteristics of specific parties involved in the transfer process discussed in the findings, where it was noted that different people have different knowledge needs and knowledge levels and may also have distinct learning abilities. A further advantage of using the HoloLens is the playful handling and interaction with the device. Prior studies regarding KES via stack overflow show that gamification contributes positively to knowledge transfer by encouraging participation in the knowledge community (Vasilescu et al., 2014). Apart from the HoloLens, a tablet is also used to visualize the set-up steps in the form of videos and to display key set-up figures in the form of set-up times and set-up paths (see Fig. 5.3). Here, a set-up process can be edited collaboratively by adapting the sequence of the individual steps.

As mentioned in Sect. 5.4.2, regular feedback mechanisms can lighten the load of the learning process (Wollersheim et al. 2015). Feedback can be provided by two main technical components. First of all, the assembly positions and tool dimensions are recorded by a 3D-camera, thus providing direct feedback upon the correct execution of a set-up step. This is important for avoiding collisions caused by incorrectly mounted tools (see Fig. 5.4).

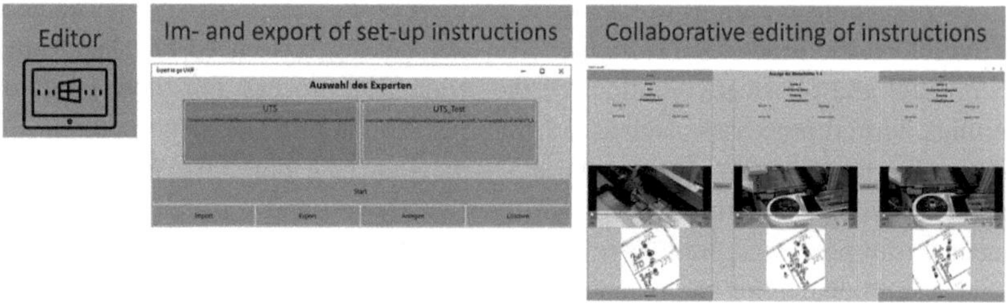

Fig. 5.3 Visualization of the set-up steps on a tablet to adapt the sequence

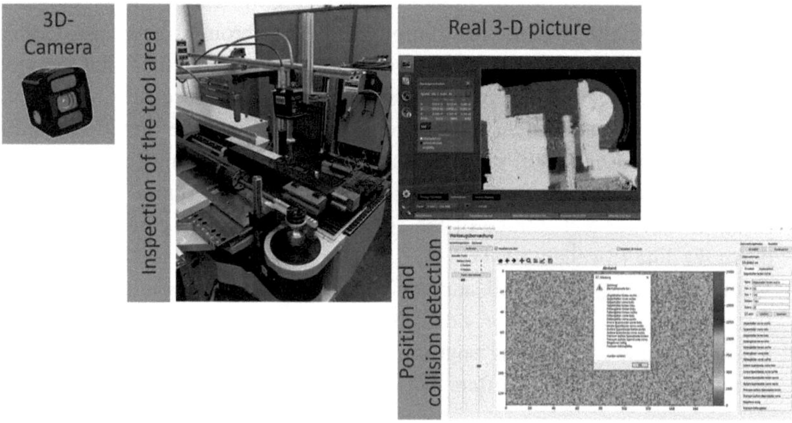

Fig. 5.4 Machine sensory equipment and the 3D camera

A second sensory recording of the set-up steps for feedback generation can be seen in Fig. 5.2 [d]. Here, the sensory equipment of the HoloLens can enable knowledge providers to record the set-up paths. An augmented visualisation of them can then be provided as feedback to assist subsequent operators. The set-up paths are also recorded on a hall plan and can be visualized on the tablet for adjustment of the set-up sequence (see Fig. 5.3).

Previous considerations of CPS have focused on just the tools and mechanical activities on the production machine. However, settings also need to relate to the theoretical and physical conditions of the production process. For these kinds of settings, virtual processes, such as calculations using a finite element method, can offer important insights into a material's behaviour (see Fig. 5.5). With the help of the results, setting instructions can then be derived and transferred to subsequent set-up processes. Similar results can be achieved by accessing the machine controls and displaying the machine parameters via an OPC-UA[8] interface. Both of these measures can influence knowledge transfer in at least two ways:

1) Aggregation of the described data creates comparability between individual set-up processes and, for the first time, hidden correlations can be depicted and a conceptualization of the settings achieved through these transferable measures. Operators are constantly on the lookout for clues that provide information about process parameters. The underlying mechanisms relating to states that experts might take to be self-evident and the direct derivation of specific measures can be further supported by visualizing (and detailing) defined and relevant process parameters. Recurrent patterns in this data can result in transfer and storage of the correlations for further conceptualization down the line (Sweller, 1994).
2) Virtual production data and augmented reality have the potential to create free experimental spaces for would-be setters where set-up activities can be rehearsed without disrupting ongoing production processes. This is hard to do at present, despite what it offers to the acquisition of embodied skills.

In summary, with the help of CPS technology, the mechanisms of expertise sharing observed in our empirical work can be supported and developed in new ways that extend beyond existing approaches to KES. The interplay of adapted visualization, sensory support and the utilization of virtual and real process data is, in our view, decisive in this regard.

Despite the possibilities offered by CPS, there are limitations. A number of limitations have been noted in the literature, mainly concerning the scalability, robustness and performance (Lee, 2008; Liu et al., 2017). As the solution put forward here only has the status of a prototype, these kinds of issues remain to be addressed. Early testing has already revealed some concerns, particularly inaccuracies in the sensory recording of the environment and resulting deviations in the

[8] The Unified Architecture (UA) from the OPC foundation: https://opcfoundation.org/about/opc-technologies/opc-ua/, accessed on 2nd November 2018.

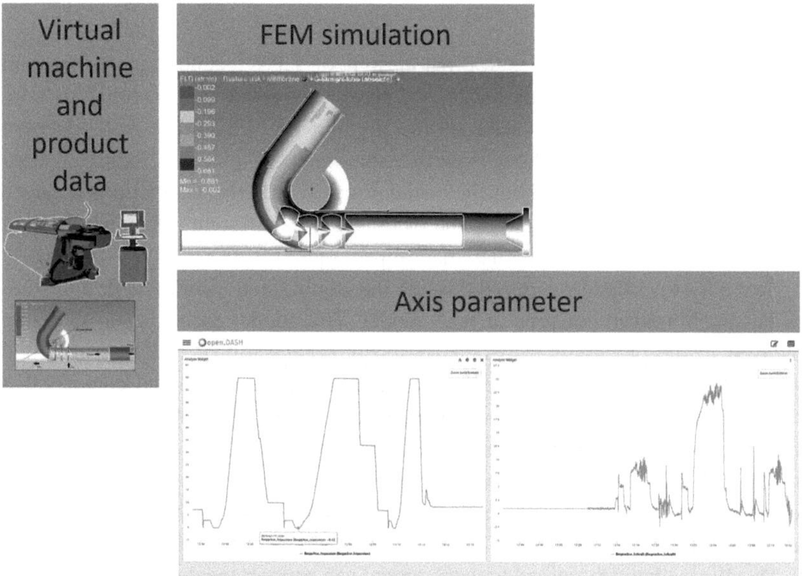

Fig. 5.5 Virtual machine data and axis parameters

presentation of the set-up instructions. Some set-up cases are not yet sufficiently accurate because of technical limitations. These are not foundational issues and will be solved over time, so they are not critical for success in the future. For some set-up processes demanding a lot of embodied knowledge, there are also some uncertainties regarding privacy. The sensory recording of set-up activities (see Fig. 5.2) by the HoloLens and the 3D camera (see Fig. 5.4) and the options for evaluating set-up data (see Fig. 5.3) with the aid of the editor make set-up data transparent and allow conclusions to be drawn about the set-up performance of the setter. An essential component of the technology is the creation of data and visualization and subsequent optimization of the underlying processes. Since the set-up process often involves highly person-dependent activities, there may be sensitivities that will need to be assessed through further study.

5.6 Conclusion

In investigating opportunities for the design of CPS technologies to support machine operators with their tasks, we have uncovered the potential to contribute to and advance the state of the art in KES. First, this work deepens existing understanding of how expertise sharing takes place in industrial contexts. It provides findings about the social structure and the interactions between actors involved in such initiatives and examines how different roles and different levels of knowledge and expertise may impact upon them.

Secondly, knowledge acquired through the ethnographic work and design studies has been used to construct a KES model. This model accounts for all actors involved in the process and associated social, organisational and environmental issues. Empirical results have been used to illustrate the situational contexts that influence KES and we have seen that different situated feedback mechanisms enable knowledge seekers to act, interact, and reflect upon knowledge being shared with them, so that they can assemble know-how. We have also seen that, in addition to the practical implementation of KES through feedback mechanisms, conceptual knowledge based on physical observations can be developed.

Thirdly, we have discussed how the model can support the design of innovative technologies to support similar problem domains. Taking into account the model elements and properties, technologies from a CPS layer can be instantiated and configured to support KES in industrial contexts. Focusing on visualization by means of AR glasses (e.g. HoloLens) and their interaction possibilities is important in industrial contexts because it responds to the need to convey knowledge that is often embedded and embodied in action, which is critical for KES in this kind of environment.

To sum up, CPS offers far-reaching opportunities to support KES. We are far from grasping and understanding all these possibilities. As argued in Sect. 5.4, we are potentially entering a third generation of KES research and can expect new developments in the near future. However, to understand the full potential of this type of technology, these types of systems need to be deployed in real settings. Only in this way can issues of privacy and ownership be properly explored and effective solutions be developed. This requires evaluation of a working system, which is clearly not what we have presented here, though our design workshops have moved us forcefully in that direction. This underscores the value of a DCS-based approach in this domain because it offers an established mechanism for not only understanding user needs and developing potential technological solutions, but also a way of studying appropriation once the solutions have been deployed. Inevitably, new issues will then emerge. It is also important to further verify and refine the model as the research progresses. The primary value of this current work is what it offers towards the concrete development of CPS-based KES solutions that are grounded in a solid body of empirical findings. Importantly, it also offers a very real way forward in tackling the problems of conveying embodied knowledge that have plagued the development of KES in its previous iterations. We would not claim to be the first to recognise this potential. What we have done is to put it all together in an application domain that is both under-studied and particularly in need of these kinds of solutions, even if significant issues still need to be resolved.

Designing Cyber-Physical Production Systems for Industrial Set-up: A Practice-Centred Approach (INTERACT 2021)

6

Industrial set-up has long been a focus of scientific research, largely because it entails substantial cost overhead for manufacturing companies. Whilst various efforts have been made to optimise this process, mainly in terms of time and other resources needed to accomplish it, to date little can be found in the HCI literature about how digital technologies can support workers who engage in it. This article sets out to addresses this gap in the literature by introducing a Design Case Study carried out for the conception of a CPPS (Cyber-physical Production System) to support machine operators with industrial set-up. Our contribution is therefore threefold: first, we describe and discuss the results of an in-depth ethnographic study, carried out under the premises of the grounded design (GD) research paradigm, to uncover practices of machine operators to inform design. Second, we introduce a series of design implications drawn from those results. Finally, we demonstrate how those design implications have informed the participatory de-sign activities pursued for the conception of the CPPS in question. In so doing, we advance the state of the art on the design of digital technologies to support people working with industrial set-up and open new research directions on the subject.

6.1 Introduction

Industrial set-up refers to a set of preparatory actions on a machine or a tool prior to the start of a production cycle (Palanisamy & Siddiqui, 2007). This is a core and time-critical operation in manufacture and in many cases without set-up there is no production. Therefore, supporting it is key to maximising efficient production and to address the challenges from the trend towards decreasing

S. Hoffmann, *Knowledge Sharing on the Shopfloor*,
https://doi.org/10.1007/978-3-658-50133-4_6

order/delivery sizes and increasing the range of produced artefact (Van Goubergen & Van Landeghem, 2002). Furthermore, the resulting need for flexibility stemming from increasing globalization, customer expectation, intense competition, as well as short innovation and product life cycles faced by either large, small or medium-sized enterprises for many years now (Vicente, 2000) has direct implications for industrial set-up. On the one hand, there is a need to reduce resource allocation and, on the other hand, there is a demand for ever more flexible production cycles and ever more varied quality demands (Shingo, 1985). This need for flexibility ramifies at a worker level too, through the demand for flexible competences to handle heterogeneous set-up processes. This may entail being able to handle a large range of different products, use a variety of mate-rials with varying degrees of acceptable tolerance, and all of this on a number of different machines. This leads to a variety of KES (Knowledge and Expertise Sharing) issues (Argote & Ingram, 2000), which can be potentially addressed by CPPS (Cyber-physical Production Systems) (see Sect. 6.5).

A review of the HCI literature demonstrate that little is known about how these systems can be designed to effectively support machine operators in industrial set-up (see Chap. 4 and 5). In view of the relevance of human-centred methods for the conception of systems that can effectively support users, this article aims to contribute towards filling or, at least, mitigating this gap. To fulfil our goals, we conducted an ethno-graphically informed investigation in four SMEs where the goals were to better under-stand the practices inherent in the set-up of bending machines and how CPPS could potentially support these practices. So, the research questions that this article sets out to answer are:

- What challenges do machine operators face in practicing industrial set-up?
- What HCI aspects must be considered in the design of CPPS, so that the resulting solution can successfully support machine operators to overcome the challenges they face in regard to industrial set-up?

Our study is oriented towards the GD research paradigm: a praxeological worldview, which highlights the importance of understanding practices for the design of useful and usable digital technologies (Rohde et al., 2017). This is a well-stablished paradigm for HCI research, increasingly being used by the community, rooted on a pragmatic approach to design research and predicated on clear scientific practices (Rohde et al., 2017; Wulf et al., 2015, 2018).

Our research questions are clearly informed by GD and go to the heart of HCI. The first research question focuses on understanding practices: according to GD, clearly understanding practices is key to identify the design space for the conception of new and innovative technologies that can effectively be used and appropriated. Designing useful and useable technologies is a central aspect of HCI since its very beginning (Benyon, 2010; Harper et al., 2008; Sharp et al., 2006). The second research question focuses specifically on identifying the HCI aspects that would contribute for the design of usable and useful CPPS to support machine operators. With this question we set out to investigate, among other things, aspects of the user interface and the interaction with it, seeking to advance the state of the art of the HCI literature in terms of designing CPPS for similar contexts and processes.

The presented findings are based on rich descriptions of the work processes involved in industrial set-up, shedding light on the practices of machine operators and the difficulties they face with such processes. They allowed us to elaborate a set-up model to unpack possible opportunities for the design of CPPS for industrial set-up. Hence, our contribution is threefold: (1) we present an in-depth user study in manufacturing contexts, which remains visibly under-addressed in HCI and CSCW, exploring the potential of CPPS to support machine operators in industrial set-up (Sect. 6.4); (2) we introduce a set of design implications, providing useful directions to follow when designing CPPS to support industrial set-up, drawing attention to important interaction challenges (Sect. 6.5); (3) we illustrate how those design implications can be applied in the design of a CPPS to support industrial set-up (Sect. 6.5).

6.2 Related Work

In this section we provide relevant background information regarding the concept of CPPS and reflect on how these systems can potentially support industrial set-up. We also examine current HCI und CSCW literature on KES, which we found to be a central aspect of industrial set-up and the challenges that machine operators face on a daily basis, as discussed in our empirical sections. It is worth noticing that this section is neither meant to cover all research studies on the subjects above, nor to deeply discuss the reviewed studies. Instead, this section is meant to setting the scene for the analytical developments of Sects. 6.4 and 6.5, which further elaborate on them.

6.2.1 Cyber-Physical Production Systems and their Potential for Supporting Industrial Set-up

CPPS consist of autonomous and cooperative elements and subsystems, e.g. Smart Machines and Smart Factories, which are connected to one another from the process level to the production level, depending on the situation or context (Monostori, 2014). Their characteristic features include the networking of various production components, such as machines or tools, as well as the data sets characterizing them (Marcus, 2015; Monostori, 2014). Concrete application examples of CPPS exist in both the industrial context and inter alia in aerospace, healthcare, civil infrastructure, logistics, military or defence, automotive, energy network and agriculture (Baheti & Gill, 2011; Rajkumar, 2012; Schuh et al., 2015).

Lee et al. (2015) introduce a structure and architecture for CPS (Cyber-Physical Systems)—and consequently CPPS, which are nothing else than CPS designed specifically for production contexts—starting with data collection, moving through analysis, up until final value creation. This architecture, according to the authors, should serve as a guideline for the implementation of such systems in the industry. These guidelines are supplemented by practical applications and techniques that underpin the theoretical architecture with practical implementation possibilities. Nevertheless, the design and development of useful and usable CPPS has proven to be challenging, which calls upon further research and development in the area.

Authors such as Lee (2008), Monostori (2014) and Paelke and Röcker (2015) have discussed some of the challenges associated with the design and implementation of CPPS. These are usually described in terms of increased technical complexity, the need for new interaction concepts that can consider the unpredictability of interaction between physical and the virtual worlds and the lack of prototyping and test tools. All of these challenges are relevant to HCI research and yet underexplored (Schuh et al., 2014). In particular, qualitative insights that explore contextual variation and the need for better tailoring of solutions are hard to find. In view of its relative novelty, this lack of research regarding CPPS is understandable. However, this is likely to change within the next 10 to 15 years, as investment in Industry 4.0 from national governments begins to kick in (Kagermann et al., 2013). Together with the potential advantages of CPPS, e.g., enhanced productivity and flexible production systems (Monostori, 2014; Paelke & Röcker, 2015; Post et al., 2017), this trend underscores the timeliness and relevance of our own contribution.

6.2.2 The Social Nature of Knowledge and its Relevance in Industrial Contexts

Knowledge is a socially constructed and often distributed asset, whose management is deemed an important strategic resource for companies and a source of competitive advantages in the market (Ackerman et al., 2013; Argote et al., 2000; Grant, 1996; Grant & Baden-Fuller, 2004; Levin, 2000; Spender, 1996a; Watson & Hewett, 2006). In CSCW, the focus on issues of knowledge has been rather more on the sharing than on the management of such an asset. CSCW researchers have long since been investigating those issues under manifold designations, e.g., organisational memory, collective memory, collective intelligence, expertise location, etc. (Ackerman et al., 2013; Randall et al., 1996) at least since the 1990s. All of these studies examine the role of information in organisational settings, concentrating on the social context and emphasising communication amongst knowledgeable workers in contexts like safety critical environments and office environments (Ackerman et al., 2013).

The HCI literature has also touched upon the matter, particularly with regard to the design of digital visualisation technologies for KES. Burkhard and Meier (2005), for instance, address issues of sharing by introducing two theoretical concepts for knowledge visualisation, based on the use of visual metaphors and knowledge maps. The authors discuss how visual metaphors are powerful for KES but yet underused in organisations. Their results demonstrate the usefulness of visual metaphors in motivating people to engage in knowledge exchanges and in supporting learning processes. In particular, the authors found visual metaphors to support the presentation of new perspectives, increase remembering, enhance the focus and concentration of the viewer, and structure and coordinate communication. Many other studies followed Burkhard and Meier's. However, most of them have focused on teaching and learning (Keller & Tergan, 2005; Patterson et al., 2014) or issues of information overload (Mengis & Eppler, 2012), rather than on KES within organisational contexts. In manufacturing contexts, even less material has been produced.

The assumption that the primary issue in relation to KES lies in making the previously tacit somehow more explicit has been challenged within the HCI and CSCW (Ackerman et al., 2013; Schmidt, 2012). Research on the field has shifted the emphasis to the social, processual and other contextual factors which may influence KES. This, effectively, can be thought of as a specific element of the 'turn to practice' which characterises much CSCW work. This re-emphasis is particularly important when the potentially transferable skills under examination are 'embodied', relying as it were on a 'feel' for things (Clarke et al., 2003).

6.3 Methodology

In this section, we provide an overview of our overall methodological approach. We will not delve into details here, as these are given in Sects. 6.4 and 6.5 ahead. We will also address the limitations of our methodology in turn, so that the readers can judge its trustworthiness and authenticity, in assessing the quality of this research (Guba, 1981).

For our own purposes, we have drawn on Wulf et al. (2015)'s DCS (Design Case Study) framework for GD (Rohde et al., 2017). GD is a solid and well-established design research paradigm, which puts focus on practices for the understanding of issues of design and the quality of designed artefacts. DCS, on the other hand, is a research design for this paradigm. Case studies, which have its origins in the social sciences, have been consistently used across different disciplines (e.g., psychology, sociology, HCI, and CSCW) to investigate and shed light on assorted types of phenomena (Bryman, 2008). DCS is built around this socio-scientific tradition and is, therefore, of highly scientific relevance.

The framework is mainly organised in three phases. The first phase sets out to understand user practices that can potentially be supported by technological solutions. The second phase, and actual design of the technology, comprises traditional user-centred and participatory design methods—see e.g., (Muller & Kuhn, 1993; Sharp et al., 2006)—to build a socio-technical system concept. The final phase encompasses a longitudinal study to investigate and document the deployment and appropriation of designed technological artefacts in the users' social systems. This article's contributions are based on activities of the first two phases of the DCS framework, namely pre-study and design. In particular, we focus on the results of the shadowing, eye-tracking, in-depth interview, document analysis, scenario-based design, DW (Design Workshop) activities and high fidelity prototyping activities. It is worth pointing out that the design phase was not constrained to scenario-based design, DW and high-fidelity prototyping; medium-fidelity prototyping, heuristic evaluation (Nielsen & Molich, 1990), cooperative evaluation and lab usability test were also involved in this phase, as seen in Fig. 6.1, although these are not ad-dressed in this article—therefore they are greyed out.

Our pre-study consisted of an in-depth ethnographic study carried out over a period of 10 months (Jun 2016—April 2017). By the end of this phase, we devised a series of initial design implication for AR-based CPPS to support machine operators with KES concerning context-specific industrial set-up, as seen in Sect. 6.5. These design implications have been based on the machine operator practices and the challenges stemming from them, as introduced in Sect. 6.4 below.

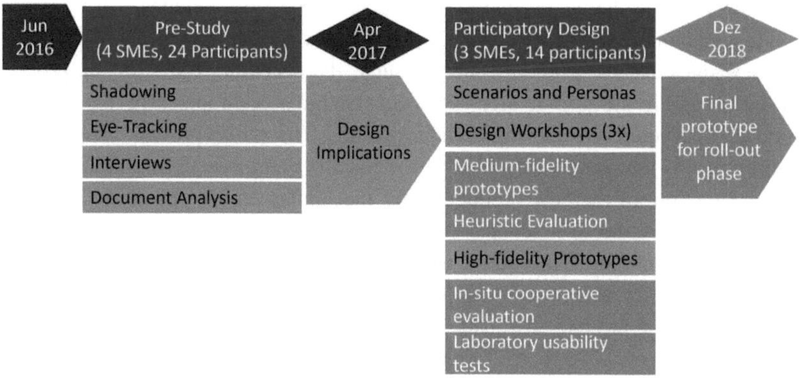

Fig. 6.1 Research Design Overview

The identified design implications informed the design activities of the second phase, which started with a series of DWs, carried out between April and August 2017. These workshops have been carried out with representative users, therefore, observing the premises of PD (Participatory Design) approaches, which entails the involvement of the participants in the design decisions across all the design phase (Muller & Kuhn, 1993). The results of the workshops fed into the prototyping activities, which resulted in the design concepts presented across Sect. 6.5, as the design implications are discussed. The design phase activities extended until December 2018, as seen in Fig. 6.1.

6.4 Uncovering User Needs Concerning Industrial Set-up Processes

In this section, we introduce our findings regarding machine operators' practices concerning industrial set-up and associated user needs stemming from it. These findings underpin the elaboration of a practice-centred model for industrial set-up as well as the design implications and envisaged solutions introduced across Sect. 6.5.

The case we examine involves the set-up of machines used to bend material like metal tubes. In order to produce different products, these machines must be equipped with different tools. Figure 6.2 shows the rough flow of the process, which is described in more detail across Sect. 6.4.1. Overall, the process starts with the machine being pre-pared for the process. This includes, for example, moving machine axes to certain set-up positions, so that parts can be removed and new ones can be installed. Subsequently, the necessary tools for the process must be found in the storage areas and organised in a desktop nearby the machine to be set up, therefore requiring certain logistics. The set-up process includes dismantling tools that were previously in the machine and assembling new ones for the manufacture of the new product. After the (dis)assembly operations, the production starts. As these machines are CNC (Computer Numerically Controlled), they must be configured through a variety of parameters. These parameters influence the efficiency and cost of the production by deter-mining its cycle time and the quality of the product. An inspection establishes the quality and, if necessary, a new iteration cycle happens. For this purpose, the error pattern is viewed and the corresponding parameter is adjusted in the CNC code.

In order to understand the user needs regarding the process, we carried out a 10-month ethnographic study. Our study included investigations in 4 SMEs in 2 European countries. The main data collection instruments were: shadowing, for in situ data collection about the participants' work practices and the social system in which they unfold; eye-tracking, for detailed information about the steps involved in the process; and in-depth semi-structured interviews, to discuss design opportunities for a CPPS support and issues arising from the shadowing and eye-tracking data. Eye-tracking sessions were recorded with both the eye-tracker cameras and a stationary video camera, giving us micro and macro representations for post hoc analysis. The in-depth interviews were audio recorded and later transcribed. We performed a total of 14 shadowing sessions, with each session also featuring at least one eye-tracking session. Interactions observed before, during and after these sessions were document-ed through fieldnotes. A total of 24 in-depth interviews ranging from 45 to 120 minutes were performed.

Overall, 24 workers across the 4 SMEs participated in the study. Out of these, 7 were from company A; 13 from company B; 2 from company C; and 2 from company D. All companies are medium-sized and produce components for various customers. The age of the participants varied from 20 to 60. They occupied different roles in the companies, as for example foreman (n = 5), production engineer (n = 3), machine operator (n = 9), process owner, etc. They also had different educational backgrounds, for instance, graduated unspecialised (n = 2), graduated specialised (n = 6), masters specialised (n = 8) and job tenures (varying from 1 to 20 years of experience). This diverse group facilitates a better representation of the different stakeholders of the system (Sharp et al., 2006).

We are aware that this is a relative small sample and that the findings of our study cannot be generalised. This is a widely acknowledged and accepted limitation of qualitative studies (Bryman, 2008). However, we have been careful to address issues of trustworthiness and authenticity (Guba, 1981) to assure the quality of our research. For that, we have used two strategies. First, we have used different data collection methods to allow for triangulation during the data analysis, i.e. cross-checking the consistency of the findings resulting from the data collected from these different methods (Bryman, 2008). Second, we have drawn on a systematic data analysis technique (Braun & Clarke, 2012), to support us in the generation of the findings, as made clear below. The triangulation performed demonstrated consistency of the findings across the different data sources, reinforcing the trustworthiness and authenticity of the findings. This becomes visible when the findings coming from the pre-study interview presented in Sect. 6.4 are corroborated by the ones coming from the DW in Sect. 6.5.

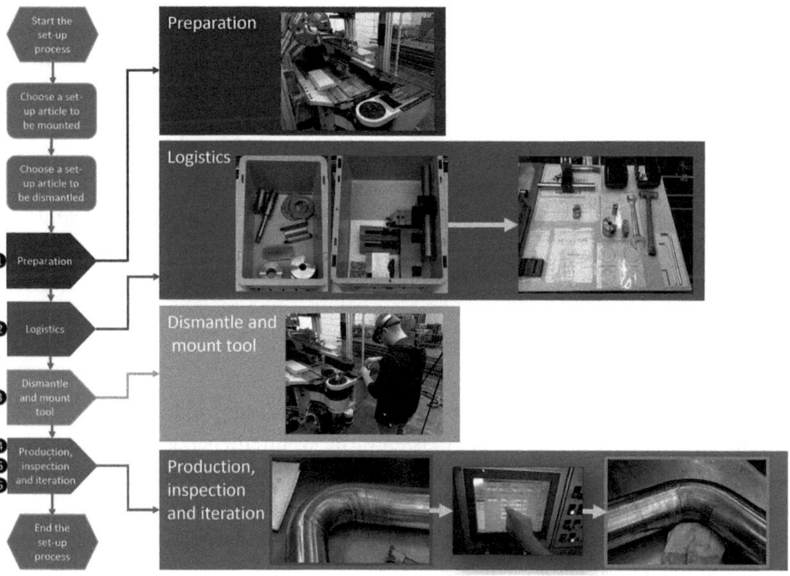

Fig. 6.2 Presentation of the general steps of a set-up process on a bending machine

The interview transcripts, fieldnotes and eye-tracking recordings were subjected to a TA (Thematic Analysis) according to Braun and Clarke approach (Braun & Clarke, 2012), which entails a set of well-established steps involving open coding of the media excerpts, systematic revision of the coded segments and the identification of code-families and their relationships, to elaborate a deep understanding of the explored contexts and/or phenomenon. In relation to the eye-tracking material, we have gone through the videos recurrently, using the Tobii Pro Lab software and have also coded the relevant video excerpts. Information about the start and end time of the excerpt was recorded in a spreadsheet and assigned a referent code. Memos about the excerpt were also written to support posterior processing of the analysis.

The pre-study data sources were coded thoroughly. More than 70 codes were identified and developed during this initial phase—e.g., sequential execution of steps, expertise-based solutions, strategies to find answer to set-up problems. These codes were further developed into themes, through careful analysis and characterisation of their relationships. Four main themes emerged from our analysis: (1) the workflow nature of industrial set-up; (2) the mixed relationship of dynamic and static elements with mechanic and non-mechanic operations; (3) the highly knowledge intensive character of the process; and the (4) potential challenges in interacting with any digital technology while working on it. We address each of these themes in turn in the next sub-sections and illustrate them with quotes from our participants. The quotes are associated with the participants who provided them. We refer to the participants through the notation (participant number, job position, company, data source), where data source refers to the data collection instrument that originated the quote—i.e., interview, shadowing, eye-tracking or DW.

6.4.1 A Workflow-like Process

Our analysis of machine operators' practices concerning industrial set-up suggested that the process can be clustered in 6 interdependent phases which resembles in many aspects a workflow. Based on the practices involved, we have named these phases as: Preparation (Phase 1), Logistic (Phase 2), Tool and machine set-up (Phase 3), Production (Phase 4), Inspection (Phase 5) and Programme iteration (Phase 6).

Workflows have been discussed in CSCW research as a sequence of subtasks in work processes that are carried out cooperatively. These subtasks are assigned to different workers, which contribute towards the accomplishment of a common goal. The route of the work is automatically defined as subtasks are completed and directed to the person responsible for the next subtask (Poltrock & Grudin, 1997). Like workflows, the industrial set-up can be split in several subtasks that will route the flow of work to the next subtask(s) upon their completion. Unlike workflows, these tasks are performed by the same worker and can happen in parallel with each other at times. This has important implications for the design of computer-aids to support it, as observable in Sect. 6.5. In the following, we discuss and illustrate each of the industrial set-up phases that we have identified through the analysis of our empirical data.

(i) *Preparation Phase*

Our observations show that industrial set-up starts with preparation activities that specify the production process and are based on existing production resources, e.g., semi-finished products and machine and personnel availability. To efficiently carry out a set-up, machine operators need clear planning guidelines regarding the set-up to be carried out *"[...] so that you have a more rational and orderly set-up" (P7, Operator, Comp. A, Interview)*. However, this is not always easily achieved:

> *On the one hand, the stock must be minimised by making the production variable and carrying out many set-up operations. Conversely, it is important to keep the overall set-up times low by means of a small number of procedures. (P4, Production Engineer, Comp. A, Interview)*

Hence, there is constant tension regarding how to respond to divergent production demands, which relates not only to the final product but also to the economics of production. We observed that participants would benefit from a working environment where there were fewer changes in production planning and therefore fewer interruptions. For the participants a lack of continuity is not just a disturbance, but is suboptimal. This results in time lost to both reconfiguring and restarting the process. Thus, the preparation phase includes a planning problem which should somehow be overcome, for example, by providing virtual process data to support planning decisions, as suggested in Sect. 6.5.2.

(ii) *Logistic Phase*

Parallel to all set-up phases, operators must deal with the logistics of the process, which refers mainly to bringing the tools and materials necessary to the set-up to the place where set-up will happen. Unsurprisingly, this is a critical step, which can potentially impact on the overall set-up time. If parts are not where they are expected to be, set-up time will increase (MacKelprang & Nair, 2010; Shingo, 1985). This is an issue concerning KES, as discussed in the design implication introduced in Sect. 6.5.4.

Even the most experienced operators have problems with logistics from time to time. Analysis of the eye-tracking records showed that logistic activities can account for up to 21% of the set-up time. In the course of the observations we noted, in particular, that additional paths followed during the set-up contributed to disturbances in an ordered set-up sequence. For instance, the set-up of the same machine, with the same tools, for the same product, by machine operators with comparable experience varied from 63 to 97 minutes (mean = 79 minutes). Closer analysis of the eye-tracking data also showed that the routes leading to increased set-up times could have been avoided if a clear assessment of logistic activities had happened at the beginning of the process. These aspects informed the elaboration of the design implication presented in Sect. 6.5.1.

(iii) *Tool and Machine Set-up Phase*

In this phase, the necessary tools and machines components from the previous production order must be removed from the machine and the new ones should be assembled. Our observations showed that much of the set-up time was invested in this phase. Uncertainties here had a serious impact on the overall set-up time and it was not uncommon for workers to draw on the knowledge of their colleagues to solve particular problems. P5 (Operator, Comp. A, Interview), for instance says: *"If I can-not solve a problem, I can call at any time or use WhatsApp and then we do that in this way"*. Indeed, the relevance of experience-based knowledge for the tools and set-up phase was a strong feature of our fieldwork data from the outset. Although it has been acknowledged in the literature that set-up procedures are highly dependent on the skills of the workers on the shop floor (Goubergen & Lockhart, 2005; Shingo, 1985), so far this has not been appropriately explored. Our own data sheds light on this issue, drawing attention to the need for a system that can support seamless KES among workers, in particular those demanding particular know-how: *"Mounting the tools, every beginner with support would be able to do this. Changing values and parameters is, in turn, a matter of experience." (P7, Operator, Comp. A, Interview)*

Hence, as the complexity of the mechanical activities increases, the know-how of the machine operators becomes decisive for a successful and efficient set-up. In principle, knowledge of the set-up process qualifies employees to perform these activities without further restrictions. As discussed in the next section, this raises relevant issues of KES, which are addressed in the design implication discussed in Sect. 6.5.4.

Overall, our findings suggest a need for instructions that generally represent the steps concerning the tool or machine set-up in question. This tool *"should actually have a representation of how the tool should be mounted on the machine" (P6, Operator, Comp. A, Interview)*, so as to ensure that the set-up runs smoothly. This also relates to the implications in Sects. 6.5.2 and 6.5.3.

(iv) *Production, Inspection and Programme Iteration Phase*

These three remaining phases have been found to be seamlessly interleaved and, therefore, are presented together here. Overall, our analysis suggests that the Production phase—when the artefacts are really manufactured—does not present any special challenges with regard to time or content concerning set-up. The Inspection phase—where optical tests and tactile measurement are carried out in a test run component—did reveal some issues, however. In the course of the actual set-up, the verification of component quality plays a special role because the results of any tests have direct implications for the set-up itself and the subsequent programme iteration phase. *"If the part is not true to gauge, I must intervene in the process and change machine parameters." (P1, Foreman, Comp. A, Interview).*

Bending processes generally provide for a gauge test by the operator. The key thing about gauge tests is that geometrical deviations are directly recognised and converted into changes to the machine programming. This programme iteration phase can be described as success-critical. It is characterised by intensive parameter adjustments on the man–machine interface (MMI) as deviations in the manufactured product are spotted. These aspects connect directly with the design implications presented in Sects. 6.5.1 and 6.5.2, as will become observable.

6.4.2 Static and yet Dynamic

Another relevant finding from our analysis is that industrial set-up involves both a static and a dynamic dimension, which have certain relationships with mechanical and non-mechanical set-up operations performed during it. Figure 6.3 introduces a model that we have elaborated out of the practices that we have observed in this regard. This model is organised into four abstracted areas of activity, concerning characteristics relating to the documentability and explicability of the set-up operations we have observed.

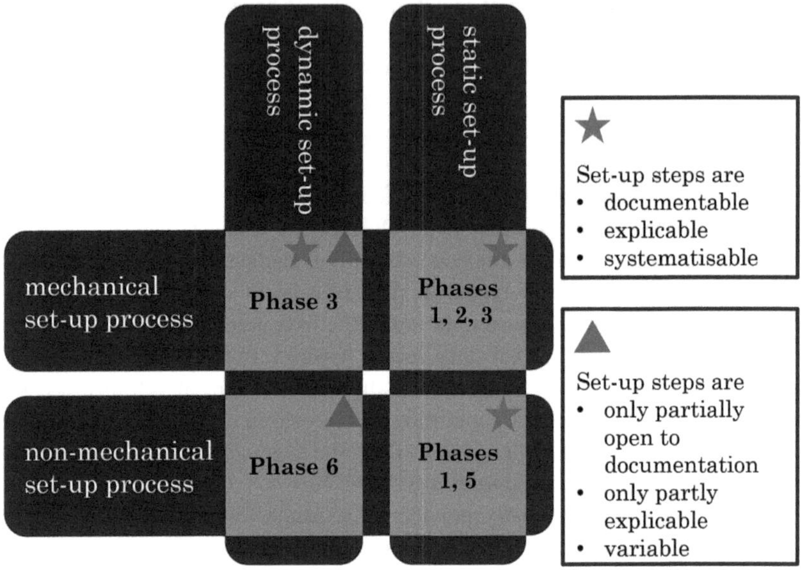

Fig. 6.3 Practice-based model for industrial set-up

From our perspective, this model represents a relevant advancement in the under-standing of industrial set-up. Existing models, like the SMED model proposed by Shingo (1985), depicts a twofold process consisting of external and internal set-up operations, our findings suggest this is a limited way of seeing industrial set-up. Our findings suggest that set-up is far more complex than it has been depicted to date. Our model therefore provides a more nuanced treatment of the practices relating to set-up.

In principle, our findings showed that it is possible to distinguish between relatively constant, or repetitive, set-up operations (static) and highly variable set-up operations (dynamic). For instance, some of the findings previously introduced demonstrate how industrial set-up refers to a rational and orderly process, which can be carefully planned. During our triangulation activities, the eye-tracking and the observation data of the logical activities showed that the Logistics phase also displays characteristics of static operations, since tools and other components required for the set-up must be placed at defined locations in order to allow easy access and quick handling. The Tool and machine set-up phase also involves some static operations, for example dismantling tool and machine components, although some of the operations from this phase would better fit the dynamic aspect of the process.

With regard to the mechanical and non-mechanical characteristics of the process, our findings suggest that the mechanical part of the set-up includes mounting and disassembly, whereas the non-mechanical part contains interactions with the user interface of the machine—e.g., configuring the CNC code. As a matter of example, the Preparation, Logistic and Tool and machine set-up phases are characterised by mechanical operations, within which the machine operator has to do physical work to complete the set-up process. The Preparation and Inspection phases have some operations which do not involve any mechanical interaction.

In terms of the characteristics of the steps involved in industrial set-up, our findings suggest that some of them are documentable. In other words, there are steps within the set-up process that can be easily documented by virtue of their simple and quick explicability. In Nonaka et al.'s (2000) words, these steps can be easily made explicit. From our own observations, these steps are also easily 'systematisable', for instance, by the use of checklists, as suggested by the SMED approach (Shingo, 1985).

Despite the fact that many set-up steps are explicable, documentable and systematisable, many of them are not, either because the machine operators cannot always articulate the reasons for their actions or because the underlying set-up actions display high variability, as illustrated by previous findings. This is particularly the case with non-mechanical operations and mechanical operations related to the dynamic dimension of our model. An example of the former are manual adjustments of the machine programme and, of the latter, adjustment of the position of a tool on the machine axis. These issues have implications, as discussed in the Sect. 6.5.5.

6.4.3 Highly Dependent on Knowledge

The findings presented so far suggest that industrial set-up is a knowledge intensive process. The high proportion of knowledge intensive operations gives KES a decisive role here. Our analysis revealed a strong need to initiate KES among colleagues because expert knowledge lies within a very restricted cohort. Moreover, as contingencies arise in this context, the knowledge needed is held by relatively few (Ackerman, 1998; Randall et al., 1996):

> *[...], there is just a lack of documentation so this mainly remains 'in the head' knowledge of the individual employees. If today three employees leave the company and tomorrow*

> *three new ones are hired, then a massive problem arises. (P4, Production Engineer, Comp. A, Interview)*

This is particularly relevant for new employees, who often lack experience. However, the knowledge in question is both extensive and diverse, so conventional KES (Ackerman et al., 2013; Nonaka, 1991; Randall et al., 1996) is especially challenging:

> *Even if everything has been shown [to you], you have to make your own experiences. There are many tricks that you do not immediately master. It is incredibly extensive what can happen there. These are many things that cannot be passed on. You can manage a large amount, but everything will never be passed on. (P5, Operator, Comp. A, Interview)*

Furthermore, personal transmission of experience can become burdensome: *"Over time, that is exhausting. I had to explain every step what he should do." (P14, Operator, Comp. B, Interview)*. There are two issues that contribute to this. First, as visible before, there is a systematic lack of documentation. Another is that the existing documentation is usually very abstract and often outdated. As a result, the existing documentation generally ends up not being used. Sharing knowledge and expertise emerged as a success-critical basis for allocation of resources—see implication for design in Sect. 6.5.2. If an effective system for KES were in place, *"I would not sit here and I would be released [to be working on something else]" (P5, see earlier, Interview)*.

Nevertheless, such a system is not easy to devise. In general, various information formats are considered helpful: *"Visual and written information and also a video" (P6, Operator, Comp. A, Interview)*. The persistence of embodied knowledge emerges as an important requirement for the tool: *"If you get help from the experienced co-workers, they make the changes and explain, but in the end, you look only and then you forget at some point" (P6, see earlier, Interview)*. This relates directly to the fact that experience-based knowledge is mostly of the embodied kind (see Chap. 4). In other words, much of the understanding of how to handle particular parts of the process only becomes visible when it is observed in action. This is difficult to convey (Nonaka, 1991; Spender, 1996b). Such difficulties call for innovative ways to record and visualise this type of knowledge. CPPS, it has been claimed, is about to result in a revolution in the way that knowledge and expertise can be shared, by providing ways to conveniently capture and display knowledge embodied in action (see Chap. 4), without the need to translate this knowledge into propositional knowledge, the driving approach in industrial and organisational management (Nonaka et al., 2000), as discussed in the design implication introduced in Sect. 6.5.1.

6.4.4 Interaction Challenges in Industrial Contexts

Set-up processes take place in real production environment contexts. Requirements for the design and development of a system support to help machine operators, need to take into account the realities of these kinds of physical environments. For instance, both our observations and interviews stressed the fact that operators must have *"both hands" (P10, Operator, Comp. B, Interview)* available to carry out set-up work. Moreover, we have observed that the workspace where industrial set-up un-folds is very limited, *"due to the large number of tools and tools combined with limited storage space, we always have a problem with space" (P5, Operator, Comp. A, Interview)*. Last but not least, *"dirt and noise" (P10, see earlier, Interview)* must also be taken into consideration. One the one hand, the system must be sturdy enough to take on eventual accidents without breaking: *"You need to have something you can work with. If it falls off, don't break it" (P11, Operator, Comp. B, Interview)*. Conversely, the environmental noise must not interfere on its functioning or in the interaction with it, as observed in our fieldwork, meaning that voice commands, at least in the current stage of the art, would be an unfeasible mechanism to interact with the envisioned technology. These aspects have informed the elaboration of the design implication discussed in Sect. 6.5.1.

6.5 Designing CPPS to Support Industrial Set-up

As previously mentioned, the findings from our pre-study led to the identification of a series of design implications for the design of CPPS to assist machine operators with industrial set-up. These implications had mainly to do with the six identified phases of the set-up process that have been elaborated out of the understanding of the machine operator practices concerning the process. We brought these implications to the attention of our fieldwork participants in a series of DWs, discussing with participants the extent to what the identified implications would correspond to their needs and expectations. In these workshops, potential technologies that could be used to address those requirements have also been discussed. The results from these activities have ultimately led to the design solutions that we introduce and thoroughly discuss in the course of this section.

In total, 3 DWs have been carried out—1 with participants from Comp. A, 1 with participants from company B and, 1 with participants from Comp. C. Identified requirements have been introduced to participants by means of scenarios and personas, following a scenario-based approach to design (Carroll, 2000). DWs lasted from 4 to 8 hours. Around 6 participants were involved in each of them. Most participants in the DWs had also participated in the pre-study, and therefore could confirm or disconfirm our interpretation of the data. DWs have been audio recorded. The audios have been transcribed and also submitted to TA, as the pre-study data, i.e., code categories have been identified and revised and the relationship between those categories have been explored elaborated and formalised in themes (Braun & Clarke, 2012).

In summary, 6 themes have been identified in terms of implications for design, namely: (1) use of sensors to collect real data; (2) use of digital simulations in situations there is no real data available; (3) support for aggregation digital data to work practices; (4) support for KES; (5) support for data configuration; (6) provision of feasible interaction in industrial contexts. These themes are not only based on the data analysis of the DW data, but also on the pre-study data.

Interestingly enough, the identified themes resonated to a large extent to with Lee et al.'s 5-level architecture for the design of CPS, introduced in our related work section (Lee et al., 2015). Nevertheless, our findings extend Lee et. al's architecture, as Lee et. al's architecture concentrates solely on the technical aspects and does not engage with the socio-technical aspects of the design and the affordance of creating an environment in which KES processes can take place. The following sub-section details how these 6 design implications can extend Lee et al.'s 5-level architecture for the design of CPS and what they meant for our design decisions (Lee et al., 2015).

6.5.1 Use Sensors to Collect Real Set-up Process Data for Subsequent Use

Our findings suggest that contextual information relevant to the process is something that can be captured, combined and prepared using a CPPS that draws on human practices or technical sensor technology. This somehow relates to Lee et al.'s smart connected level of CPPS (Lee et al., 2015). Appropriate sensor technology could be implemented through different identification systems, like barcodes or RFID systems (Finkenzeller & Waddington, 2003) and connected to the CPPS. In so doing, the system would be able to support machine operators with operations from phases 1 and 2 introduced before:

The Holo-Lens would be connected via the W-LAN. That means I can access everything that is available via W-LAN, i.e., if we have such a reader for an RFID chip that we can access via the network. (P23, Comp. B, DW2)

This design implication led to a series of design decisions for the AR-based CPPS elaborated during the participatory activities carried out during the design phase of our DCS. Figure 6.4 illustrate how this have been implemented in practice.

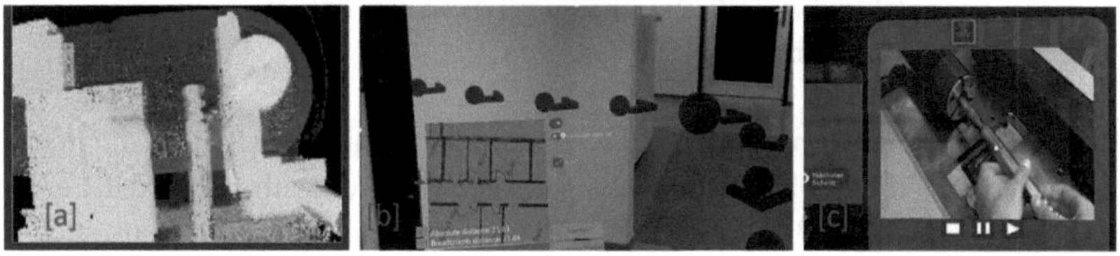

Fig. 6.4 First level: Sensory check of the size and position of the mounted tools [a], the recording of the set-up paths [b] and the video recording of the set-up interaction [c]

In summary, for logistic activities, sensor technologies have been employed. Sensor technologies can play an important role in recording paths and providing data for subsequent optimisation (see Fig. 6.4 [b]). As evident in the quote below, this sort of tracking would be a relevant aid in optimising the time used for the set-up process:

From the methodological point of view this is a typical SMED procedure [...] (T)here is a so-called spaghetti diagram where the paths covered by setter during this set-up process are shown. (P25, Comp. B, DW2)

In addition to that, machine operators can be assisted in unambiguously identifying the tools as well as in determining their position on the machine axis by selecting the tools through sensor-based recognition (see Fig. 6.4 [a]). Not only that, by means of sensor technology, local knowledge embodied in action can potentially be recorded in videos and shared (see Fig. 6.4 [c]). In other words, sensor technology has the potential to show product-specific characteristics to workers in real time, enabling context-sensitive dispositions, so to avoid wasting time and other resources in the accomplishment of the set-up process.

Real machine data can also be beneficial if individual values can be assigned to a quality image of the article. This speaks directly to user needs that were captured in phases 5 and 6. Furthermore, there is a need to share knowledge about the machine settings that could be used for different situations, i.e., the knowledge about the different parameters that can be defined for each tool axis, which describe the geometry of the article to be manufactured and the procedural. Capturing and sharing this knowledge in a contextualised way can offer a significant added value for other operators in the course of the dynamic set-up process.

6.5.2 Provide Virtual Process Data to Support Decisions in Situations where Real Process Data is not Available

The results we have presented across this article implies that CPPS should not only be consisted of real data, but should also be enriched with virtual data, as for example, by blending 3D holograms with the real machines as demonstrated in Fig. 6.5 [a], or by providing simulation data as seen in Fig. 6.5 [b]. This resonates with Lee et al.'s Conversion level of CPPS architecture, which suggests that CPPS should generate smart analytics for machine component health, multidimensional data correlation and degradation and performance prediction. Phase 3 introduces requirements for a virtual confirmation of the assembly process.

In this case, a virtual construction environment can potentially circumscribe this activity [51, 52]. Separately, the assembly process must be free of collisions. This can be ensured through virtual kinematic simulations in the preparation phase and with the help of AR visualisations on the real machine during the set-up process (Fig. 6.5 [a]). In addition to that, phase 6 presents an even more complex requirement for virtual data concerning the configuration of the machine parameters. Hence, virtual production data, which considers the material reactions of the semi-finished products in detail, can offer some added value, because the settings can be derived in advance of the real set-up process. At the same time, the need for specific settings can be clarified by means of material stress parameters (Fig. 6.5 [b]). Here an operator can interpret the colour coding of the places with the highest stress and introducing specific countermeasures, locating the exact position of maximum load and tracing it back to a tool movement with the aid of the diagram.

The virtual generation of process data, then, can allow for the translation of certain aspects of local knowledge into propositional content, to be used during the performance of actual mechanical processes. This virtual data can potentially complement the real data generated by the use of sensors.

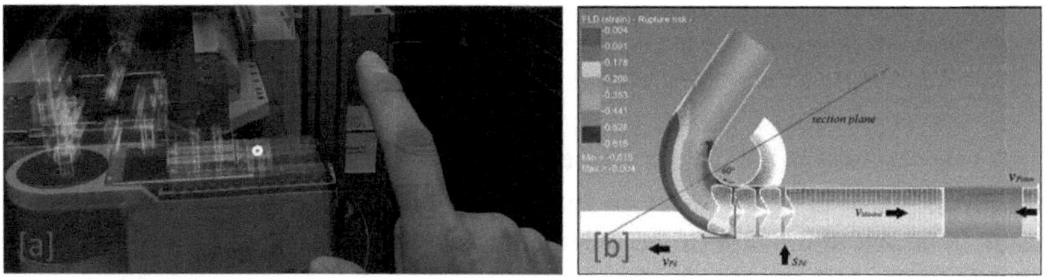

Fig. 6.5 Second level: The recording of the virtual set-up interaction [a], the generation of virtual process data by means of an FEM simulation [b]

6.5.3 Supporting Human-centred Aggregation of Data

From our results, it is relevant to support a human-centred aggregation of data, or in other words the aggregation of experience-based knowledge 'owned' by workers. Put differently, data has to provide an added value for KES:

> *If I first use this system to create a kind of story book for the set-up process, to link individual data to a manual and then afterwards have the possibility to move or change this data again in a different order in my office, then from my point of view I have all the possibilities I need to create a knowledge base. (P25, see earlier, DW2)*

However, this is limited by the capacity of people to recall all of the specific details (Argote et al., 2000; Sweller, 1994). For instance, the demands from phases 1 and 2 represent the need for data aggregation in the form of (primarily) master data about an article and its associated tools. The master data serves as a cornerstone of an article to be produced. Likewise, they form the foundation for their digital representation (Tantik & Anderl, 2016). This resonates with the third level of Lee et al.'s architecture for CPPS, which has to do with a Cyber level where twin models for components and machines are generated and managed. The twin model is a virtual representation of the real machine and describes the behavior of the machine in the virtual world.

Our results also indicate that a set-up operation can benefit from a digital representation of the haptic aspects of the process (see Fig. 6.3 'static set-up process') as well as essential data for the machine's adjustment (see Fig. 6.3 'dynamic set-up process').

> *If I have the 6 different positions in front of me [where I can assembly a tool] in the machine, the HoloLens must show me the position with a virtual representation of the*

tools. In addition, information about the assembly is stored in a window above the machine. All information must be directly visible when looking at the machine, but must not obscure the assembly location. (P31, Comp. C, DW3)

Figure 6.6 [a] and [b] show the implementation. The data recorded on the lower two levels are assigned to a set-up step and saved afterwards. The virtual data is stored in an AR animation of the set-up interaction and machine data is visualised on a dashboard. However, the data only adds value for a learning-friendly environment if it is aggregated realistically and is situationally relevant, detailed and problem-oriented (Lee et al., 2015). Furthermore, the data must come directly from the production process. A reduction of complexity can be achieved: by a) ensuring the provision of the data takes place in a context-sensitive fashion and b) that it is directed to support key elements of decision-making (Schuh et al., 2015). The aggregation of real and virtual data and their representation in embodied action and through propositional knowledge correlate strongly with the mechanical and non-mechanical as well as the static and dynamic set-up components.

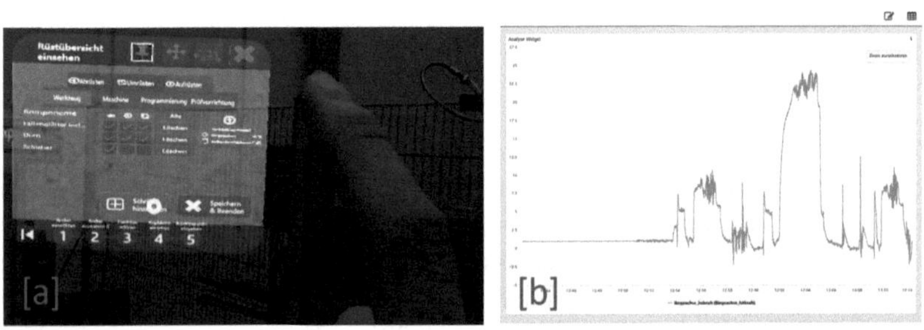

Fig. 6.6 Third level: The storage of the recorded set-up data in the HoloLens [a], the visualisation of the machine data [b]

6.5.4 Support Knowledge and Expertise Sharing

Integrated simulation, connected systems, remote visualisation for human and collaborative diagnostic and decision making are some issues addressed by the fourth level of Lee et al.'s (2015) CPS architecture, namely Cognition level. Not surprisingly, the long-term storage of knowledge and the individual and independent accessibility of multimedia content is an important advantage in an industrial environment (Caudell & Mizell, 1992; Goh, 2002; Roberts, 2000; Watson & Hewett, 2006). This has been extensively discussed within the CSCW and HCI literature, which has been investigating OM (Organisational Memory) and KES issues since the early 1990s (Ackerman et al., 2013).

As discussed across Sect. 6.4, some critical operations of the set-up process are intensively influenced by experience-based knowledge. Nevertheless, we have observed that written documentation is fragmentary, not up-to-date and does not capture the necessary level of detail:

> *I think it would be cool if we could manage to get the mediation that you can say, I get for the respective tools that I want to install now, with which torque they theoretically have to be tightened and which screws, for example, or all of which must be used for this tool. That I really only have to grab my belt and I have everything I need. I also know with which pretension. (P23, see earlier, DW2)*

Advanced systems have significant potential to support the contextuality information (Paelke & Röcker, 2015). Chapter 4 introduces a discussion about how CPPS can support KES by allowing new ways to convey information embedded in embodied action. We have further pursued the initial ideas and have made it concrete in the technological aid that we put together to support machine operators in carrying out machine set-up. Figure 6.7 shows the visualisation of the set-up interaction, which was recorded on the second level of the CPS architecture and now is aggregated on the fourth level to a complete set-up instruction, using an AR visualisation. On the basis of this interactive instruction, set-up information is available independently of people's availability and a KES can take place on the basis of the recorded and visualised set-up information.

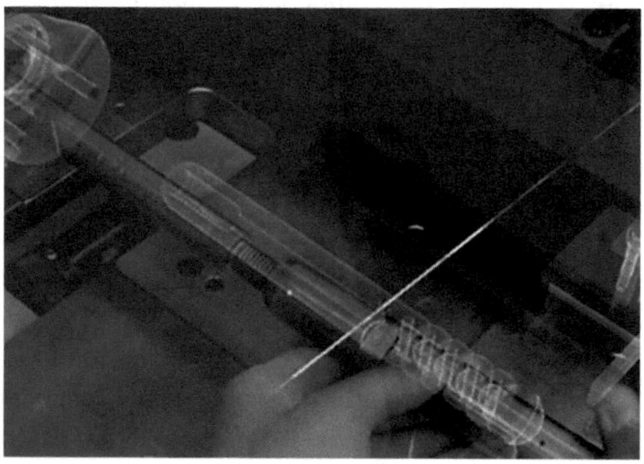

Fig. 6.7 Fourth level: The replay of the recorded set-up data

6.5.5 Support Knowledge and Data Configuration

The findings concerning the phase 1 stress the importance of standardised processes and therefore the need to provide interactive checklists according to well-defined and clear foundations in order to avoid failures. During the DWs, it was made clear that for supporting machine operators with their daily set-up activities, the system has to be able to access databases that contain logistic real-time information with regard to the name, storage location and condition of the respective target object:

> *So, there are databases for certain things and links for certain things, so I know which tools are necessary for this article [...] it would make sense for our set-up editor to continue thinking in databases, i.e., we create a complex tool database. (P7, Operator, Comp. A, DW1)*

In a second step, users can potentially interact with the system by confirming or verifying physical availability as well as the target state of the target object. Moreover, by such means, users can be guided step-by-step through the working task and receive multimedia support in terms of texts, pictures and videos (Fillatreau et al., 2013):

> *If I can see a 3D model, then that is certainly not bad. Additionally, it can be enhanced with pictures and videos. With this data a standard set-up instruction can be created [...] probably not perfect yet [...] Because at the beginning the experience knowledge is not there yet, it is only built up with time. (P33, Comp. C, DW3)*

The observations above resonate with the fifth Lee et al.'s (2015) architecture for CPPS, namely Configuration level. Iterative approaches where machine operators evaluate a best practice, then modify accompanying processes such as logistic tasks, influence the actual set-up tasks. Although, some mechanical set-up operations are static, it is possible that there will be an impact on the handling processes as well. Existing set-up instructions need to offer adaption options for these changes.

This implication resonates with the realisation that frameworks, which actively support learning processes of the worker, must adapt to innovations and changes (Vicente, 2000). Overall, then, any support system will need to meet these requirements. In particular, changes in set-up order need to be developed cooperatively and, if appropriate, through actual interaction. For this purpose, a tablet client, which can adjust the aggregated set-up data by changing the sequence of the set-up steps, was implemented (see Fig. 6.8).

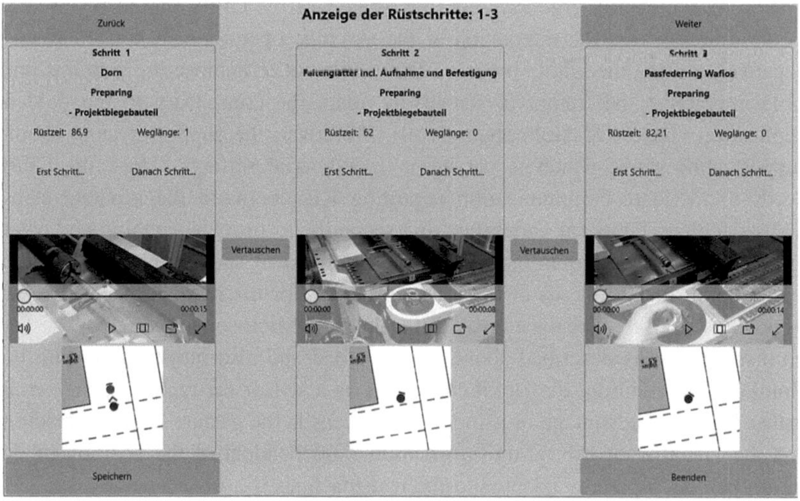

Fig. 6.8 Fifth level: Adaptation of the set-up process

6.5.6 Provide Feasible Interaction Mechanisms for the Manufacturing Context

A final implication for design that we have identified concerns a vertical level that permeates all other 5 levels of Lee et al.' (2015) architecture: it concerns the interaction challenges faced in the everyday life activities of machine operators. These challenges mainly emerge from the interactions between users and the system in the environment where the system is deployed:

> *[...] weight of the glasses, [...] the wearing comfort [should be] there. The recording quality of the videos has to be good, that you don't have to focus so exactly on the window, but that you can move freely. The same with the sound quality. When I record something, I want to make sure that it is of the appropriate quality, that the background noise is not so present and that you can understand what the colleague is saying [...] (P30, Comp. C, DW3)*

In more detail, these challenges encompass: 1) allowing the machine operator to operate hands-free; 2) local, mobile and decentralised access to information because of the manual reconfiguration processes which are performed directly at the machine; 3) recognising the limited working space around the machine.

Therefore, the central ergonomic design aspects of the CPPS should be oriented to the principle that the interaction of the machine operator will not be affected negatively during the whole process. The aim is not to achieve the simplest possible interaction, but rather to sensitively adapt the complexity to the field of application (Janlert & Stolterman, 2010). Moreover, the support system should support shift work, which is pervasive in industrial settings. Thus, the CPPS needs to facilitate communication regarding KES between the separate steps, firstly, to offset temporal variability, and secondly, to provide teaching and training material. These requirements cannot be limited to specific phases, but they are essential for all phases of the set-up process. For the reasons listed here, we decided to use AR-visualisation with the help of Microsoft HoloLens. The set-up instructions, as described above, are recorded and also played back with the HoloLens. In addition, the HoloLens serves as a sensor for recording the set-up paths. The main advantage of using the HoloLens is the gesture-based interaction and the visibility of the set-up information—independent of the position of the set-up person—spatially adapted to the real machine.

6.6 Conclusion

This article introduces a series of timely and relevant contributions to the HCI research. Timely because they refer to the design of CPSS, a type of system that has been receiving increasingly attention of the community and the market, but has not yet been fully explored in the HCI literature, especially because only recently it has been proposed as a solid concept to be explored in design (Lee et al., 2014; Monostori, 2014); relevant, because they address a context not yet widely explored in HCI (manufacturing contexts) and introduce findings shedding light on how these systems should be designed, in order to be useful and usable and to fit to users practices. These three last aspects are key for system acceptability and appropriation, as discussed extensively in the literature (Wulf et al., 2015).

In answering our first research question, we have demonstrated that industrial set-up is a knowledge intensive process involving an ecology of practices that make it really challenging. Our findings certainly overlap with other CSCW studies on KES, as for instance the case of sharing knowledge about technical questions in organisations (Ackerman, 1998; Randall et al., 1996) or about medical diagnosis and care activities in medical settings (Bardram & Bossen, 2005; Strauss et al., 1985). Nevertheless, our findings point towards a case of knowledge concentrated in the hands—or the heads, if one will—of very few actors, where the 'social distribution of expertise' is of a highly concentrated kind. Different from Bardram and Bossen (2005), who discuss how medical and care knowledge was distributed across different actors and its implication for mobility work, our case reports on a small number of experienced machine operators that hold important knowledge concerning the set-up of the varied machines. The manufacturing context explored also has its particularities, as demonstrated by our findings, as has the addressed process, i.e., industrial set-up. All those elements sum to the novelty of our contribution. Furthermore, our implications for CPPS, which have not yet been extensively explored for KES, provide us with opportunities to advance the state of the art.

In terms of our second research question, our findings suggest that CPPS should be able to capture and support both the static and dynamic dimensions of set-up, ultimately leading to more effective KES by the machine operators. The use of sensors can potentially support operators in capturing different set-up scenarios that can then be learnt from down the line. Furthermore, the subdivision into processual knowledge which is represented by embodied action, as well as the presence of propositional knowledge can be reflected in the representation. While the static and mechanical set-up process is primarily characterized by embodied action components, the dynamic and non-mechanical set-up process is characterized by propositional knowledge components. Furthermore, as discussed in Chap. 4, CPPS can support effective KES by providing new ways to capture, process and visualize relevant knowledge that can potentially trigger improvements in the process.

All in all, the findings regarding manufacturing contexts (e.g., re. noise in the environment) can potentially be transferred to other similar contexts (e.g., construction contexts). The same applies for the findings concerning processes which are workflow like and demand the use of both hands to be accomplished. The observations and the solutions herein presented can, we argue, inspire other design solutions or even sparkle new HCI research.

Producing and Consuming Instructional Material in Manufacturing Contexts: Evaluation of an AR-based Cyber-Physical Production System for Supporting Knowledge and Expertise Sharing (CSCW 2022)

7

Fast-paced knowledge and expertise sharing (KES) is a typical demand in contemporary workplaces due to dynamic markets and ever-changing work practices. Past and current computer supported cooperative work (CSCW) research has long been investigating how computer technologies can support people with KES. Recent claims have asserted that augmented reality- (AR-)based cyber-physical production systems (CPPS) are poised to bring significant changes in the ways that KES unfolds in manufacturing contexts. This paper scrutinises such claims by implementing a short-term evaluation of an AR-based CPPS and assessing how it can potentially support (1) the generation of AR content by experienced production workers and (2) the visualisation and processing of such content by novice workers. We, therefore, contribute a user study to the CSCW community that sheds light on the use of a particular type of AR-based CPPS for KES in industrial contexts.

7.1 Introduction

The growing complexity of services and products and the rapid changes in markets have increasingly required workers to keep up to date with changes in the workplace, increasing demands that they learn new processes and practices at a fast pace (see Chap. 4). Knowledge and expertise sharing (KES) is arguably a key aspect of addressing these complexities; therefore, it requires careful consideration (see Chap. 5). In the fields of human-computer interaction (HCI) and computer supported cooperative work (CSCW), researchers have focused considerable attention through the years on (1) how KES is—or can be—mediated by

S. Hoffmann, *Knowledge Sharing on the Shopfloor*,
https://doi.org/10.1007/978-3-658-50133-4_7

computer technologies, and (2) how KES is—or can be—embedded in workers' everyday practices (Ackerman et al., 2013). In manufacturing contexts (see for example Fig. 7.1), research from these fields has recently started paying attention to the design and representation of assembly instructions through augmented reality- (AR-)based systems to overcome limitations from current approaches, which are heavily based on user manuals and standard operation instructions: see, e.g. (Agrawala et al., 2003; Schnotz & Bannert, 2003; Siltanen et al., 2007; Westerfield et al., 2015; Yuan et al., 2008). In particular, AR-based cyber-physical systems, which in manufacturing contexts have been termed cyber-physical production systems (CPPS), have been receiving increasing attention within HCI and CSCW research. These systems are expected to revolutionise how KES can be performed (see Chap. 4).

Fig. 7.1 Set-up for a bending machine consisting of twelve tools and components to be mounted on the machine (red markings) and an additional camera sensor (green marking)

CPPS refers to physical networked and engineered manufacturing systems interwoven with computational systems. These systems monitor, control and coordinate operations through the simultaneous provision and use of Internet-enabled data access and data process services (Monostori et al., 2016; Rajkumar et al., 2010). The concept of CPPS goes beyond the union between the cyber and physical worlds, spanning the interaction between their physical and computation components (Lee & Seshia, 2015).

Chapter 4 introduced a discussion on how AR-based CPPS technology can support experienced operators in authoring instructional material for fellow colleagues. This material is created by recording the assembly steps and annotating them in the system. This approach allows novice workers to easily access this knowledge. The authors argued that AR-based CPPS is on the cusp of bringing significant changes in the ways that knowledge and expertise are documented and shared between knowledgeable actors. Particularly, they offered the prospect of providing support for sharing quite specific forms of embodied knowledge. Chapter 5 elaborated on this argument and introduced a model for the design of CPPS for KES in manufacturing contexts. However, studies are lacking from real production environments that evaluate the interaction of (1) AR content created by company experts, (2) the use of AR content by novices in the course of a training process and (3) the support of the training process by sensory data. Therefore, we aim to address this literature gap by evaluating how a specific type of AR-based CPPS KES system—based on holograms, demonstration videos, sensor-based information, and 3D cameras—performs when machine operators use it to share relevant information about machine set-up. For this purpose, the paper has exclusively addressed AR content created in an authoring environment by experienced workers and passed on (unedited) to novices. We attempt to reproduce KES practices currently in place in some companies. Such practices rely heavily on producing expert-worker-produced paper-based documentation, which novice workers use as they familiarise themselves with the processes and procedures concerning machine set-up (see Chap. 4). The following research questions underpin the contribution:

- How can an AR-based CPPS (like the one explored in our study) support workers in creating and consuming set-up instructions?
- How can such systems support workers with improving their task performance?

By asking these questions, we focus on both the opportunities and limitations that such technological support may offer. Naturally, these are complex questions that cannot be completely answered through short-term studies, such as the one that we have carried out for this particular contribution. Therefore, we recommend exercising caution with the results presented, especially because they cannot be generalised. The small sample size in this experimental study and the inherent limitations of the qualitative part of our mixed methods approach prevent any attempt to generalise the results. Nevertheless, the findings that we present already illustrate relevant aspects of those questions' answers. Our findings demonstrate that AR-based CPPS can indeed facilitate creating artefacts to support KES, conveying not only propositional knowledge but also action-embodied knowledge. In this context, propositional-knowledge components are less of a challenge because they encompass knowledge that can be fully recorded and described (Brödner, 2007). Embodied knowledge describes knowledge that is anchored in a person through personality, creativity, intelligence and perception (Fitzpatrick, 2003). Thus, embodied action represents the knowledge that is more hidden in people's movements and interactions (see Chap. 4).

Our findings demonstrate that authoring environments should allow the creation of three levels of information. When using the authored content, improvements should be allowed over time to eventually build a generally accepted standard. In addition, fast visual feedback mechanisms are beneficial for understanding the instructions. Sensor values both contribute to improving the instructions and allowing the building of feedback mechanisms. Furthermore, a performance enhancement usually results indirectly: for example, reducing the time that more experienced workers must invest in guiding less experienced colleagues.

In this paper, we specifically address CPPS support of KES in the context of machine set-up. Machine set-up corresponds to a series of preparatory actions on a machine or a tool prior to starting production. This process typically occurs in a serial production between the end of one particular article's production and the start of another (Van Goubergen & Van Landeghem, 2002). Figure 7.1 shows an example of a typical set-up task. In manufacturing, this key process must be accomplished in due time; in many cases, without set-up, production is not possible. Recent research has found this process to be knowledge-intensive (see Chap. 4).

We conducted the evaluation using a mixed methods approach (Bryman, 2008; Creswell, 2014). As such, the results presented come from a two-stage study. First, the referred CPPS was tested in a natural setting in four companies with a total of about 30 hours of evaluation. Next, an experimental study followed in a controlled environment with 16 participants. The qualitative analysis carried out with the data collected in the first study informed the elaboration of the hypothesis, which has been tested based on quantitative data collected in the second study. Qualitative data collected in the second study further informed the analysis started with the first study's data. Accordingly, we elaborate on the answers to our research questions. We, therefore, contribute a study on the potential of AR-based CPPS for KES in industrial settings. Our investigation responds to recent calls for research studies targeting the use of those systems in such contexts (see Chap. 4) and more elaborate investigations that test such systems in a natural environment under real-world conditions (Dey et al., 2018).

The remainder of this chapter is organised as follows: The next section presents the related work on the use of AR visualisations in learning environments and the potential of AR-based CPPS for KES. The third section introduces the research context and provides relevant information on the design and implementation of the AR-based CPPS evaluated in this study. We then present our study's methodology and follow with introducing and analysing our findings. The chapter ends with a discussion and conclusion, reflecting on the extent to which the findings corroborate the literature's claims and the new issues they introduce.

7.2 Related Work

The potential of AR-based technology for learning has already been highlighted in past and current HCI and CSCW research (Novick & Ward, 2006; Randall et al., 1998). More recently, the potential of AR-based CPPS for KES has aroused the interest of researchers from these fields. This section refers to these contributions and briefly discusses how the AR-based CPPS solution evaluated for this paper departs from them.

7.2.1 AR-based Visualisations as Instructional Aids in Manufacturing Contexts

Classic manuals and standard operating instructions are widely used in the course of industrial activities because they can be standardised. However, these instruments can have a negative impact on actual work processes, as they are often cumbersome and make locating relevant information quite challenging (Norman, 1998). Past and current research has demonstrated the potential of AR technology in manufacturing contexts, especially in regard to guiding workers in accomplishing tasks such as assembling 3D-structure parts or following the flow of material through manufacturing facilities (Baird & Barfield, 1999; Bhattacharya & Winer, 2019; Dey et al., 2018).

Baird and Barfield (1999), for example, found that AR-based instructions would be more effective for supporting operators with manual assembly tasks. The authors ran a comparative test using four types of instructional media: (1) paper manual; (2) computer-aided; (3) opaque augmented reality display; and (4) see-through augmented reality displays. The results of correlation analysis (considering the assembly time and the number of errors made during the testing tasks) have suggested that the augmented reality conditions were more effective in guiding the participants through the test. Despite the variables that might have influenced the study's results (for example, instructions were presented on separate sheets in the paper conditions and on different screens in the computer-aided conditions), evidence indicated that the faster assembly times could be associated with the AR condition.

Similar results have been found in more recent research. Henderson and Feiner (2011) investigated the benefits of AR technology in guiding mechanics through maintenance and repair tasks. More specifically, the authors investigated the extent to which AR applications support workers with tasks concerning the installation and removal of fasteners, indicator lights and connecting cables positioned in a cramped interior. Their findings showed that the AR application used in the study allowed workers to locate the parts and carry out the maintenance faster. Likewise, Fiorentino et al. (2014) found similar results suggesting that AR-based instructions have a significant impact on workers' overall maintenance-task execution time and the number of errors committed. Echtler et al.'s (2004) study showed that a welding gun modified with AR instructions for position determination supported the user in controlling the positions for the welding spot more quickly. The advantage of this application is that the markup and the welding could take place in parallel (Echtler et al., 2004).

Many other studies have shown similar outcomes (Dey et al., 2018). One potential reason for better performance with AR support is that multimodal feedback (e.g. visual plus auditory or tactile feedback) can be more effective than visual feedback alone (Burke et al., 2006). In turn, Tang et al. provided evidence that AR technology can reduce cognitive load and, therefore, the possibility of overload (Tang et al., 2003). Feedback given by AR systems differs partly from classical systems intending to adjust the action. Digital technology provides constant, real-time updates on progress (Goomas, 2012; Hermsen et al., 2016), enabling more effective immediate feedback (Goomas et al., 2011). Research has shown that feedback on the state of a machine, for example, can be processed particularly well if the feedback and the positioning of the information move with the machine (Klinker et al., 2004).

Evidence has also shown that AR technology can increase accuracy and, consequently, reduce workers' cognitive load (Tang et al., 2003). As several authors have argued, inaccuracies may raise excessive extraneous cognitive load that can lead to overload within a restricted working memory system (Bresciani & Eppler, 2009; Paas et al., 2003). AR can potentially minimise complexity by adjusting reality to virtual representations and eliminating the need for spatial placement.

Despite the promising results of many and various studies, some researchers have pointed out AR's limitations. Regarding accuracy, Kosslyn has drawn attention to the lack of unquestionable results regarding AR-based visualisation's impact on the accurate positioning of the visualised elements (Kosslyn, 2006). Studies like Werrlich et al.'s (2019) have also checked some claims concerning the potential support of AR for assembly workers. In their study, the authors' results suggested that in some situations, assembly workers would complete assembly tasks much faster with paper instructions. Nevertheless, the authors did find that the number of errors committed when using AR support was lower than when using paper-based instructions (Werrlich et al., 2019). Blattgerste et al. showed similar results (Blattgerste et al., 2017), contrasting with the results of Funk et al. (2016), who found evidence that AR-based instructions presented on a head-mounted display (HMD) could be associated with a higher number of errors and significantly higher cognitive stress. The same applies to Büttner et al.'s study (Büttner et al., 2016). The results showed that AR with an HMD performed worse than projection-based AR—and paper. The reason was attributed to the lack of technical hardware equipment at that time. The types of hologram visualisation also influence performance. Recent research has shown that concrete in situ holograms positively influence performance (Blattgerste et al., 2018).

In the industrial environment, projection-based support systems are increasingly used. These systems can improve classic instructions because they guide users without prior knowledge through the assembly task (Rodriguez et al., 2015). In addition to the results that these systems perform better than HMD (Büttner et al., 2016), recent research has shown that the appropriation of operating complex machines is supported by projected in situ videos, thus blurring the boundary between explicit and implicit knowledge. A further advantage of projection mapping is that it supports appropriation accompanied by other people (Ludwig et al., 2019). In addition, the projection mapping's necessary sensory equipment allows a direct check of the assembly results and, thus, promotes the users' well-being

(Ojer et al., 2020). However, more critical results from studies with in situ projections have also been reported. These have indicated not all study participants improved their performance with this support and that the exact causes remained unclear (Korn et al., 2013).

Results have shown that an AR-supported learning process has no significant advantages over conventional instructions regarding the number of errors, execution time and workload. Results from previous studies have suggested that the learning curve flattens after the initial training phase (Wilschut et al., 2019). In a study on training analysis, Büttner et al. (2020) also showed that personal training (one-on-one training with an expert), training based on an in situ projection and training based on a paper instruction performed equally well in the short term. Neither training cycles nor training times differed significantly. The second training day showed different results, with personal training outperforming the other modes of training. The application of the learned content after 24 hours showed an identical picture. Here, too, personal training was significantly better. However, this effect was no longer noticeable after one week. Using AR in a learning environment to explain complex machines has proven worthwhile, as the machines can be fully explained via the new interface, and the handling can be instructed. An interface for data transfer between AR visualisation technology and the machine enables the user to see the machine's values directly in the AR interface and, thus, better understand the machine (Eder et al., 2020; Jasche & Ludwig, 2020).

Therefore, room for further research on the theme still emerges, especially in more natural settings and under real-world conditions, which can expose workers to issues other than those in controlled environments (Dey et al., 2018). Furthermore, the rise of CPPS has introduced new issues yet to be explored (see Chap. 4 and 5). In CPPS, AR technology is just part of a broader construct that integrates sensor technology and goes beyond instructions to span manufacturing processes' logistics and the quality control of resulting products. We set out to explore such complexities and argue that by shedding light on an underexplored area, we are providing an incremental yet relevant contribution to the CSCW community.

7.2.2 AR-based Authoring Tools

One demand in manufacturing contexts is to keep documentation of work processes and practices up to date (Radkowski et al., 2015). Past and current research, however, has found that such documentation is usually outdated. A possible reason is that work practices and processes evolve with the time and availability of new resources, and the production of new documentation does not follow the pace (Novick & Ward, 2006; Randall et al., 1998).

Concerning AR-based content, the literature has found that its production can be challenging, expensive and time consuming (Bhattacharya & Winer, 2019). As early as 1998, Neumann and Majoros argued for elaborating an AR-media language to facilitate the creation and deployment of AR content across platforms. According to the authors, content should be built as media that could be visualised in different contexts. This must represent geometric data in the form of in the form of computer-aided design (CAD) models (Neumann & Majoros, 1998). More recent research has shown that in an environment of complex tasks, holograms are the best fit for KES (Kosslyn, 2006; Speier et al., 2003). They can account for the position of information to be presented and the displayed object's orientation (Newell et al., 2014; Rzayev et al., 2018).

Existing authoring tools focus on transferring first-design prototypes from paper or plasticine into AR content (MacIntyre et al., 2005; Nebeling et al., 2018; Nebeling & Madier, 2019; Nebeling & Speicher, 2018). Thus, new applications have been trying to support the generation of AR content from simple 2D sketches and plastically modelled objects without having relevant programming knowledge (Nebeling et al., 2018). Nevertheless, challenges still arise with enabling end-users in their normal work environment to create and share AR content without repeated interruptions—which may affect task performance (Nebeling & Speicher, 2018; Speier et al., 1997). Hence, the increasing attention focused on new approaches to create technical documentation comes as no surprise (Bhattacharya & Winer, 2019).

The most conventional approach arguably consists of collecting summaries of existing instructions in a knowledge base over a long period. These summaries are posteriorly edited using a GUI editor and displayed in an AR view using object-recognition algorithms (Jo et al., 2014). Another traditional approach is manually creating instructions via 2D authoring wizards, which can be posteriorly displayed in a mixed-reality environment (Zauner et al., 2003). More recent approaches have tackled creating complex AR instructions based on existing 2D documentation using 3D models (Mohr et al., 2015). The proximity to 3D modelling tools is deliberately sought to make the authoring environment as intuitive as possible (Seichter et al., 2008). In addition, sensory values and complex recognition algorithms can be used to derive AR instructions when performing a task. Afterwards, they must be manually refined before novices can use the instructions (Agresti & Finlay, 2008).

In particular, the literature has shown that intelligent authoring environments have advantages in executing assembly tasks. In addition to the pure generation of virtual content, ontologies and logic are stored in the authoring environment, allowing for intelligent guidance and feedback on the performed task (Westerfield et al., 2015). Authoring environments implemented via a desktop application are preferred in certain stationary scenarios, as users perceive them as more performant (Gonzalez et al., 2019). In such scenarios, tangible AR authoring interfaces are also listed as an intuitive environment for creating AR content (Lee et al., 2004).

Using sensory information and automating parts of the AR-content creation aligns with a CPPS-based approach to KES. CPPS have great potential to ensure the networking between production plants with product design, supply chain management (SCM) and quality assurance processes through the provision of interfaces to the production plant and the collection of machine and sensor data (Wagner, 2018). In terms of KES, researchers have argued that AR-based CPPS are about to bring significant changes in the ways that knowledge and expertise are documented and shared between knowledgeable actors because they can support sharing specific forms of embodied knowledge. Whilst sensors and other types of technologies and computations can potentially record information about specific actions often overlooked, AR technology can mix this information with the real world to convey a richer representation of specific processes (Leit et al., 2016; Pacaux-Lemoine et al., 2018; Wang et al., 2014). In particular, they can offer new ways for upskilling workers by facilitating the creation of instructional material, which can be recurrently used in various situations (see Chap. 4). Additionally, sensory equipment detects failures immediately, allowing for timely intervention (Funk et al., 2017). As Ludwig et al. (2017) showed, the sensor captures within CPPS are also suitable for capturing contexts and environmental variables, thus providing a better understanding of production processes. Our study's results corroborate some of these claims, as we discuss in the next sections.

7.2.3 Knowledge-based Manufacturing

The findings of research on computer integrated manufacturing (CIM) have shown that the role of humans in computer-assisted production is still important. CIM has been extended by arguing that human skills, experience and competencies remain essential components in manufacturing. In the course of this, a 'human-centred CIM' was argued for (Badham, 1991; Brödner, 2007). The current literature around modern cyber-physical systems also argues that human work should be supported by enhancing people's physical and cognitive capabilities.

Ultimately, an interplay of humans and technology should build up in a so-called socio-technical system (Cimini et al., 2020). The goal must be to better integrate humans into production systems, using their knowledge and skills, and to gain advantages through human cognitive abilities. In doing so, human abilities are augmented. The approach is via seamless data, information, knowledge, and decision flow (Emmanouilidis et al., 2019). The latest industrial revolution no longer focuses solely on increasing productivity but rather on increasing innovative capacity through knowledge and expertise sharing. The concept of organisational memory (OM) is essential, as it shows how organisations use and maintain knowledge. Formal approaches are less successful because knowledge is neither easily stored nor easily accessible (Ackerman, 1998; Ackerman & Malone, 1990). Rather, it enables living practices and integrates knowledge creation into everyday life, which has implications for the design of assistive technologies, especially in an industrial context (Bannon & Kuutti, 1996).

7.2.4 CSCW and KES

KES has been a recurring topic of interest in CSCW since the 1980s when terms like 'organisational memory', 'organisational remembering', 'collective intelligence' and 'expertise location' flourished (Ackerman et al., 2013; Randall et al., 1996). First, the focus was on developing systems to support creating an 'organisational memory' that anyone could access when needed (Ackerman, 1998). Second, attention was drawn to the fact that, whilst OM implies a particular and static conceptualisation of 'skills', 'knowledge' and 'memory', an individual must understand the social context in which the 'remembering' process occurs (as in accessing information and using it in some way). This understanding is key to grasping what is transpiring when KES-related activities are undertaken. Therefore, communication issues related to KES and the vehicles whereby knowledge might be shared amongst knowledgeable actors became commonplace (Ackerman et al., 2013).

Moreover, in response to the growing interest in tacit knowledge in OM and information systems (IS) literature, researchers have created a debate in the CSCW literature about the distinctiveness of the practical skills and knowledge people are drawing upon when undertaking certain kinds of tasks (Ackerman et al., 2013; Schmidt, 2012). This debate is relevant for our interest in production processes because it addresses how a range of production activities should be performed, represented and shared with others who might want to engage in similar activities.

A present interest of CSCW-related design is the effect of knowledge sharing on environments where technical tools play an increasingly important role and where novice or less experienced workers use and reuse knowledge and expertise captured from experts as often as required. However, few studies have actively set out to investigate how knowledge sharing is accomplished in real-world practice. Randall et al.'s (1996) work regarding OM might be considered one notable study in this regard, where observations of one particular individual's critical role in articulating 'local knowledge' were heavily drawn upon. Another key study was undertaken at Xerox PARC regarding the sharing of information between service technicians (Yamauchi et al., 2003), which was premised on an earlier study by Orr (1996). The contribution we offer in this article advances the subject's state of the art by presenting a mature system, which has been designed and developed following well-established human-centred approaches. Moreover, this system has been initially tested to verify to what extent it can support workers with KES in industrial contexts.

7.3 Research Context and Implementation

The findings we present in this paper are based on evaluating an AR-based CPPS, which has been designed as part of a design case study (Wulf et al., 2015): a framework adopting a human-centred design approach organised into three phases. The first phase sets out to understand user practices that can potentially be supported by technological solutions. The second phase, and actual technology design, comprises traditional user-centred and participatory design methods. The final phase encompasses a longitudinal study to investigate and document the deployment and appropriation of designed technological artefacts in users' social systems. Thus, the extent to which the proposed solution meets user requirements can be evaluated (Wulf et al., 2015). With this work, we partly contribute to the third phase, as we only present the results of a short-term evaluation. The implementation is based on the key findings from the preliminary study that KES between experienced and inexperienced machine operators must be supported. The architecture's implementation is shown in Fig. 7.2.

Fig. 7.2 Architecture of the implementation

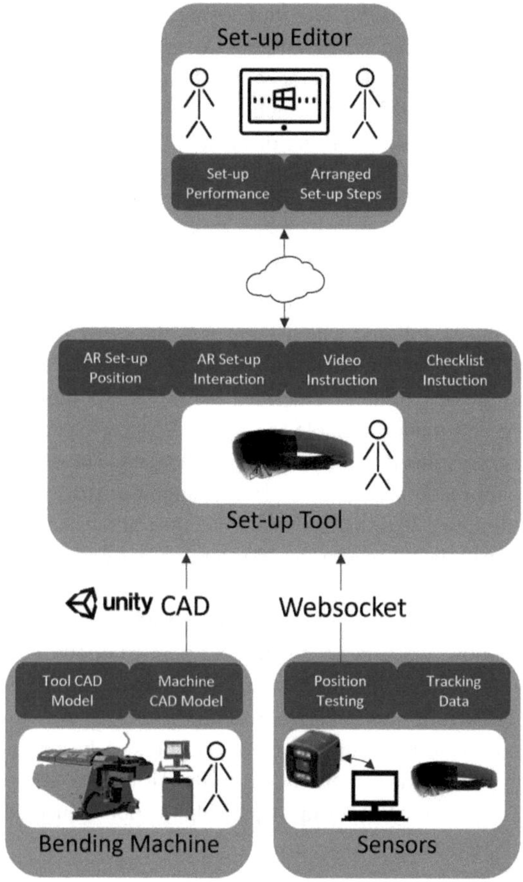

The tool and its position on the machine are specified by the CAD model and are loaded into Unity with the structure contained in the CAD model. In addition, machine components that can also be part of a set-up process are loaded into Unity. After the CAD data has been processed in Unity, the application must be recompiled and installed on the HoloLens. Afterwards, the data is available on the HoloLens application. Data from the camera sensor is processed in a camera computer. The computer calculates whether an error arises during assembly (wrong position or wrong size of a tool) and sends this error message to the HoloLens via a web socket connection. The camera is a time-of-flight camera that measures distance based on a time-of-flight measurement. The resolution is 352 × 264 pixels. The monitoring software is implemented in such a way that each pixel can be monitored. A target distance value is assigned to the pixel, and a permissible tolerance for the distance value is defined. The tolerance was defined at ± 0.5 mm, which is also approximately the measurement tolerance of the camera at a working distance of < 1 m. The HoloLens and the camera computer are coupled via a Wi-Fi connection. The error message from the camera computer is processed on the HoloLens, and the corresponding holograms of the tool components and machine parts are colour highlighted. The 3D-camera image is, therefore, neither further processed nor displayed on the HoloLens. Instead, only an error message from the camera computer is transferred as a JSON file to the HoloLens, which then uses the error message to colour-code the corresponding hologram. The CAD data is processed and combined into an instruction using the workflow described below. The complete instruction is saved as a zip file and made available via a cloud-based exchange directory. The set-up editor accesses this directory and loads the data. The adjustments made in the editor are again made available in the exchange directory and can thus be accessed by the HoloLens. The technical architecture allows experienced machine operators to independently record the instructions and pass them on to inexperienced machine operators.

As mentioned before, the set-up editor has been designed and implemented to allow workers to create set-up instructions and visualise them. The set-up editor comprises three modes: Writing, Reading and Configure (see Fig. 7.3). The writing mode allows workers to create new instructions, edit existing ones and create a camera check of the tools to be mounted. Reading allows workers to visualise the instructions on the HoloLens. The configuration mode allows a rearrangement of set-up steps with a tablet connected via a cloud to the HoloLens.

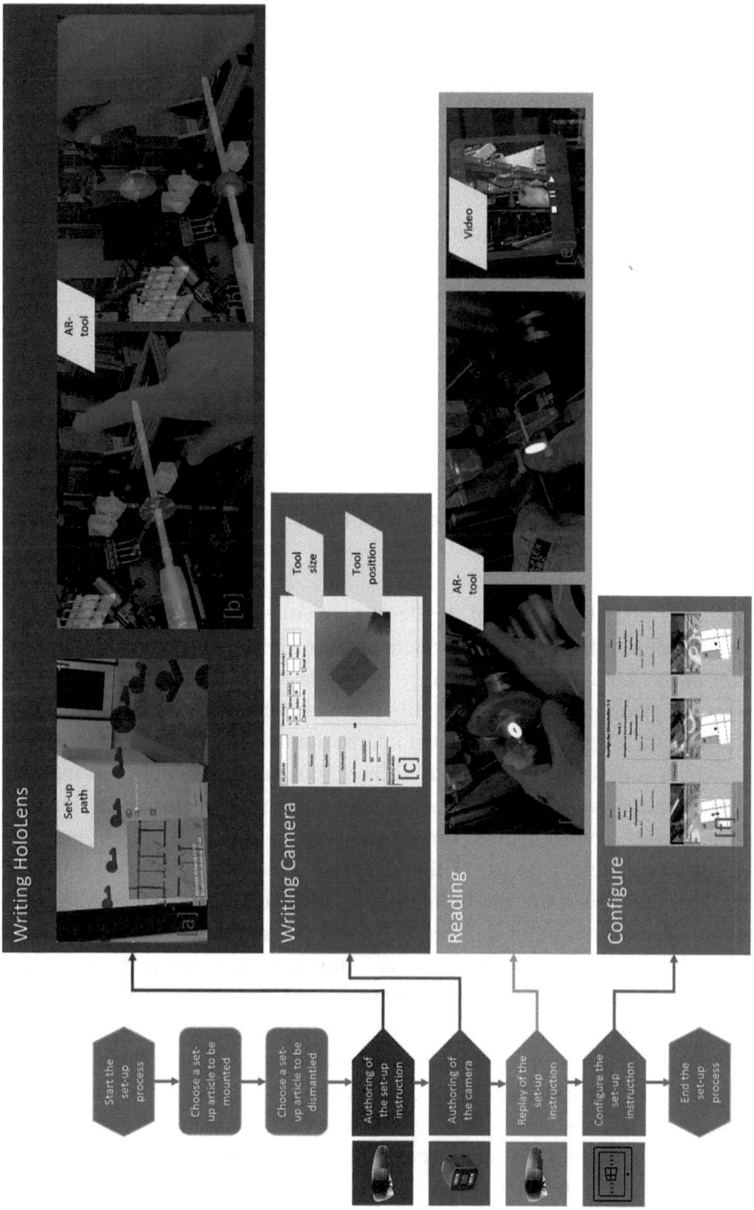

Fig. 7.3 Illustration of the set-up support workflow and implementation

The Microsoft HoloLens has been selected as the visualisation tool for the solution. The primary reason behind this choice is that the HoloLens would respond effectively to a series of design implications that emerged from the ethnographic study. This equipment would

- allow for AR visualisations to improve the comparison between the target and actual state;
- be mobile and retain all instructions and information within sight;
- provide resources for workers to take pictures and videos;
- include integrated sensors to explore the logistics of the set-up process;
- utilise external sensors for checking the results of mechanically performed set-up steps and integrate the signals of the external sensors in the field of view;
- allow for voice commands or air taps, which are key for machine operators who either need to have their hands free whilst working or interacting with the system with dirty hands;
- enable the instructions' recording and playback with the one and same hardware.

Worth noting is that at the time of the design activities, the HoloLens 2 was not available. Therefore, the evaluation presented addresses a solution using the first generation of the Microsoft mixed reality smart glasses.

Conversely, essential design implications can only be implemented with suitable AR hardware. Therefore, an adapted interaction design of the application was required, which is described in more detail here. Basically, the entire AR interaction design is based on the gesture control of the HoloLens with tap interaction. Alternatively, voice control can be used, but this proved problematic in preliminary tests on noisy factory floors. Within the HoloLens application, two different holograms are used. First are the holograms for displaying the tool positions on the machine. These holograms can be marked in the authoring mode with a tap and animated with a tap and hold. In replay mode, no further interactions with the tool holograms are implemented. Window-like holograms are used to display the different menus: e.g. for selecting what tool should be set up or the textual information on a set-up step. Basically, these holograms provide three different display modes to ensure that the user can always see the content without overlaying important areas on the machine. Therefore, the window-like holograms can be set to (1) stay fixed in one place, (2) always follow the user and stay in the field of view when the user is moving, or (3) attach to the surface. The user is free to decide on the form of display. The disadvantage here is that the user can lose sight of the fixed holograms and not find them again. In addition, the preliminary study showed that some functions should always be accessible to the user (e.g. a certain set-up overview must always be accessible). Therefore, we implemented a hologram foot menu, which is always in front of the user's feet to ensure that certain functions are permanently accessible. This feature does not disturb the user with an undesired overlay of the hologram on the workspace.

7.3.1 Authoring of the Set-up Instructions and the Camera

In order to generate documentation, operators need to start the writing mode and go through the wizard to complete the machine set-up. The wizard steps were defined according to the practices observed during an ethnographic study. The actual instruction for a set-up step consists of three information visualisations. The first is the holographic representation of both the set-up position and interaction. With the help of the holographic representation, the exact tool assembly position on the machine is determined. The operator selects the component to be mounted from the tool's holographic representation using an air tap (see Fig. 7.3 [b]). The user accesses a predefined tool list and its 3D representation. Once the position has been determined, the operator must define the assembly interaction. A coordinate system appears. Along the system's axes, the operator can tap and hold the direction from which the component must be mounted. Subsequently, an animation is automatically created, which holographically moves the components into the mounting position. If a tool consists of several connected components (for example, a tool is mounted with several screws and washers), then all components are animated according to their mounting orientation. However, since the simple animation does not always provide sufficient accuracy of the assembly interaction, the operator can also save an assembly video. The HoloLens camera takes the assembly process from an ego perspective and assigns this video to the set-up step.

Furthermore, the HoloLens and its ability to orient itself spatially are used as sensors for the set-up process. For this purpose, a path is recorded and assigned to a set-up step. The underlying methodology is derived from the set-up optimisation SMED and is applied analogously in the form of spaghetti diagrams (Novick & Ward, 2006). On the one hand, the recorded paths are represented via a drawing of the respective premises. On the other hand, they are visualised utilising breadcrumb holograms (see Fig. 7.3 [a]). The breadcrumbs additionally contain a heatmap functionality. This functionality has been implemented so that the diameter of the breadcrumb spheres increases as operators spend more time in a particular position. These data can be posteriorly analysed to investigate how to enhance a particular set-up's logistics. In addition, the HoloLens enables validating and optimising the expert-recorded set-up sequence based on the path's sensory recording, which is an essential component of direct feedback and positively affects the performance and acceptance of the instruction (Westerfield et al., 2015).

The camera provides another feedback mechanism. For this information, the tool size and position must be recorded in the authoring environment with the help of the 3D camera and position control. Additionally, a size control must be set in the self-developed camera software (see Fig. 7.3 [c]). Because the camera software is operated on a separate camera PC, the camera must be configured separately. The configuration takes place at the end of the HoloLens authoring process, as the camera can only be configured when all the tools have been mounted. Integrating direct feedback through sensors ensures that important correlations are learned more quickly (Paas et al., 2003). This is particularly critical for set-up processes, as the individual set-up steps build on each other. In addition, the sensory equipment is designed to allow reflection in action (Bryman, 2008).

7.3.2 Replaying Set-up Instructions

Once a set-up process is recorded, operators can draw on the recorded data to guide them throughout a set-up activity. As with the writing mode, the reading mode basically consists of a wizard. Operators navigate through the steps, which usually offer operators checklists to support them remembering all necessary tools and materials for the set-up process at hand. Operators also find instructions on how to (dis)assemble particular set-up components, play any audio or visual material created for a particular step, and (re)play animated holograms created during the documentation's elaboration (see Fig. 7.3 [d], [e]). Furthermore, the 3D camera mounted over the machine set-up area (see also Fig. 7.1) can accurately track the position of the installed tools. In case of a part's incorrect positioning, the system issues a warning message informing the operator. These capabilities support early identification of set-up problems that would result in wasted material, as previously, workers would have to first run a test.

7.3.3 Configuring Set-up Instructions

The set-up process is fully captured using the technologies presented and can be exported on a tablet client for analysis (see Fig. 7.3 [f]). The tablet application and has two essential functions. First, set-up steps can be analysed by examining the video files under lean aspects. Specifically, the video files are analysed for non-value-added activities to eliminate the identified wasteful activities afterwards. Second, the set-up times and distances of the individual set-up steps are displayed and also visualised on a map. On this basis, the set-up steps can now be rearranged. Afterwards, the adjusted set-up steps are exported to the HoloLens.

We have implemented this functionality as a tablet application because, in this case, several people must view the set-up contents requiring a type of multi-user environment. The tablet client is not used during the set-up process but in a downstream analysis phase, for example, in SMED workshops (Shingo, 1985). Consequently, a screen-based presentation and a form of interaction in which several people can edit the content on one piece of hardware is suitable. Nevertheless, it should be a mobile solution, as SMED workshops are often conducted directly at workplaces.

7.3.4 Resulting Workflow

The individual elements of CPPS are embedded in a set-up workflow for the set-up editor, which is summarised in Fig. 7.4. The user of the set-up support first determines an article due for production and for which the machine must therefore be set up. In addition, the article that has been produced and whose tools must be removed is determined. Both articles result in a specific set-up sequence, which the operator must create. To accomplish this, they use the HoloLens application's writing function. The function includes recording the track times for the logistic activities. For this purpose, the operator collects all the tools required and marks them on a list, recording the paths accordingly. Afterwards, the operator processes all set-up steps and creates a set-up data record for each step that can include any or all of the following elements: an AR animation, an image and a video. The step is then saved, and a new set-up step is defined. After the set-up recording is completed, the tools' dimensions and their position on the machine are recorded with a 3D camera, and the data is stored on a camera computer.

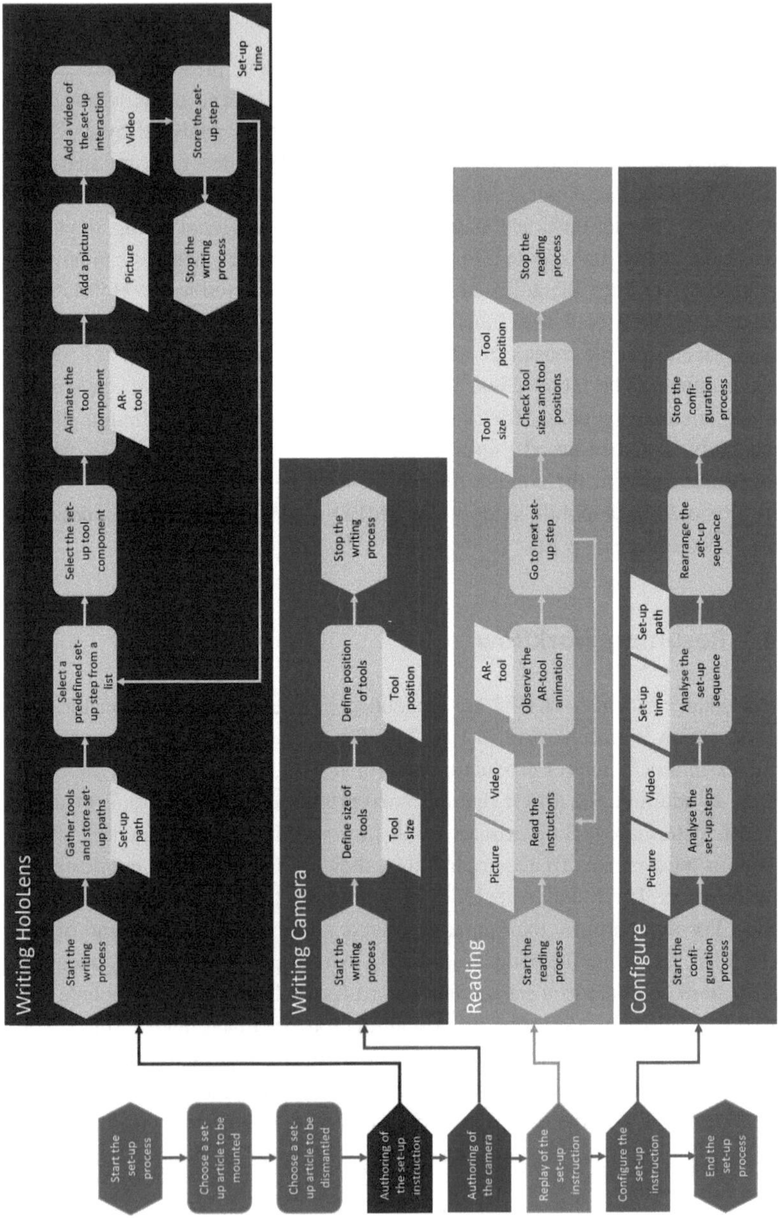

Fig. 7.4 Set-up workflow

The replay of the saved set-up steps is done with the help of the HoloLens application's reading mode. The stored set-up data is made available to the set-up operator step by step. At the end of the set-up process, the tool size and position on the machine are checked with the camera.

If configuring the set-up data is necessary, the sequence of the set-up steps can be adjusted using the tablet application. To do this, we imported the set-up data generated with the HoloLens (video, image, distances, and set-up times) for each set-up step to the tablet. Thus, a new, optimised set-up sequence can be determined and saved according to the data. The data is then imported back to the HoloLens.

7.4 Methodology and Materials

As we mentioned earlier, the goals of our study were to scrutinise an AR-based CPPS to (1) assess how these systems can support KES processes in industrial contexts and (2) investigate how the authored data can improve machine operators' task performance. The research questions that we aimed to answer with the findings presented ahead are as follows:

- How can AR-based CPPS, like the one explored in our study, support workers in creating and consuming set-up instructions?
- How can such systems support workers with improving their task performance?

In order to shed light on those questions, we conducted a two-step evaluation study on the AR-based CPPS that we have been designing within the design case study described in Sect. 7.3. To answer the first research question, we conducted the Phase 1 study. The Phase 2 study was conducted to answer the second research question. The first phase of our study was carried out on the set-up editor in natural and uncontrolled settings, as Fig. 7.5 shows.

The results of these tests were addressed in a short re-design cycle until we could establish a stable version of our prototype. This stable version was evaluated in a series of laboratory tests, where an entire CPPS infrastructure, including a 3D camera as a sensor, could be configured and tested. Unfortunately, setting up this infrastructure in the companies was not possible, as it would affect their productive activities.

Fig. 7.5 Evaluation of the implemented set-up system

Both phases of our study followed a mixed methods approach, i.e. both qualitative and quantitative data have been collected for analysis and are complementary (Bryman, 2008; Creswell, 2014). Each phase was predicated using cooperative evaluation, which is a well-established method for testing interactive systems (Monk et al., 1993). As such, the phases were planned and carried out according to Monk et al.'s (1993) run time guide.

In both phases, a first-generation HoloLens equipped with our prototype was used. Two video cameras were mounted on two tripods positioned in strategic locations to capture the users' external interactions and reactions as they completed the tasks. A lapel microphone connected to a portable audio recorder recorded the users' comments after the system's evaluation. Paper notepads and pens were used to take note of any insightful comments and impressions of the interaction.

For each phase, we completed two steps: first, we elaborated a task sheet, which featured tasks that could be performed with the system to be tested; second, we prepared a checklist for the materials necessary during the test. The checklist was completed the evening before each test was performed to guarantee that all materials would be prepared for the session. We elaborated on specific methodological aspects of each phase of our study at the beginning of the sections dedicated to their results.

Worth noting is that our methodology is situated within socio-informatics and primarily informed by the grounded design research paradigm. This paradigm is a praxeological worldview based on the assumption that, in order to design useful, usable, efficient and effective systems, one needs to engage in understanding users' practices accurately and work closely with the users to devise solutions that support them with their practices (Stevens et al., 2018). Design case studies, as Rohde and colleagues (Rohde et al., 2017) suggested, are a framework that can support design researchers and practitioners in implementing grounded design projects. A practice-centred approach to the design of KES solutions fully aligns with what has been recommended within CSCW, as Ackerman and colleagues (Ackerman et al., 2013) have thoroughly discussed. Indeed, the contributions on the subject have always paid careful attention to the socio-technical aspects, as evident in the related work presented in Sect. 7.2.4. Grounded design, in particular, brings together what has been learned for the first and second waves of HCI. It articulates the use of well-established research methods to understand user contexts and the evaluation of computer technologies. Each occurs in naturalistic (as in partially) controlled settings—usually associated with the first wave of HCI. Employing participatory approaches focuses on sharing and learning within communities of practice and is commonly associated with the second wave of HCI, which has influenced the emergence and establishment of the CSCW field (Bannon, 1995; Schmidt & Bannon, 1992). The third wave focuses on the design of individual experiences (Bødker, 2015). As such, ground design responds to what Bødker (2015) identified as a need to balance issues of the three waves of HCI.

7.5 Phase 1: Evaluation in Naturalistic Settings

In the first phase of our evaluation, we brought the system to the companies and had operators interact with the system whilst working to answer the first research question. The focus of the study is the support of knowledge transfer between experienced operators and novices with the help of an AR-based CPPS. Therefore, two scenarios were tested in the first phase: one featuring the HoloLens's writing mode (see Fig. 7.4 for an overview of the steps involved) and another focusing on its reading mode (see also, Figure 7.4 for the steps involved) combined with the configuration of the set-up steps. Participants were assigned to a particular test scenario (writing or reading), depending on their experience, qualifications and company role. More experienced and qualified people were assigned the writing mode scenario, whilst novice workers were assigned the reading mode scenario, according to the original purposes of the modes.

7.5.1 Study Design

We carried out the first phase of our study in four different companies. For this purpose, a set-up process consisting of a mounting and a dismantling process was defined with the company's Operations Management department, and the corresponding data was prepared. This data described the article to be produced (e.g. article number, article description, article characteristics and, optionally, an article drawing). In addition to article data and data relating to the individual set-up steps (e.g. the tool's name, required assembly aids, required fasteners), the CAD data for the respective tools were also prepared and made available on the HoloLens. For confidentiality reasons, we cannot disclose all the individual steps of the set-up processes, but we summarise the overall scope of the set-up processes addressed in Table 7.1.

Table 7.1 Study Design Phase 1

#	Machine	Number of set-up steps	Duration of evaluation
Company A P2, P3, P4, P6	Bending machine for small diameter	24	2 hours for writing 2 hours for reading and configuration
Company B P9, P10, P13, P14	Bending machine for large diameter	12	3 hours for writing 7 hours for reading and configuration
Company C P27	Special machine for the production of flexible metal hoses	> 90	5 hours for writing 4 hours for reading and configuration
Company D P30, P31, P32, P33, P34, P35	Press for stamping metal strip	12	4 hours for writing 3 hours for reading and configuration

The set-up process in Company C (see also, Figure 7.5, Phase 1, left picture) was extremely complex and consisted of more than 90 steps that required implementation on a fully automated machine consisting of several modules. The set-up process was characterised by tiny tool components, some of which were difficult to access, and high spatial distribution of the individual set-up steps. The set-up processes in Company B and Company D (see also, Figure 7.5, Phase 1, right and middle pictures) were carried out on different machines but shared similar characteristics. Different from the set-up process of Company C, the number of tools necessary for the set-up process in Companies 2 and 4 was smaller. The individual tools were also considerably larger and difficult to handle. The tools were spatially concentrated and most easily accessible. In Company D, a medium number of mainly small tools were spatially concentrated and had to be mounted on a bending machine.

7.5.2 Participants

In Phase 1, our target group consisted of machine operators working with the set-up process. In Phase 2, we added students who had never carried out a set-up process, as we were interested in performance aspects concerning the use of AR technology in novice users conducting set-up operations.

In total, we recruited 18 participants for the first phase of the study. All were males, as no females on the shop floor were responsible for machine set-up. Eight expert participants (with more than three years of professional experience) engaged with the writing mode scenario, and 10 participants with little or no experience of machine set-up engaged with the reading and configuration mode scenarios. Machine operators were invited to participate in the evaluation by members of the Operations Management department of the companies involved in the study. Included in the invitation to participate, the companies received the information sheet about the study. These materials enabled the companies to make an informed decision about their participation. Four small-medium enterprises from two European countries were involved in the study. Participants had different educational backgrounds and various levels of experience—some working as few as three years in the area and others working as many as 40 years. The company workers did not receive any compensation or incentives for participating.

7.5.3 Procedure

In Phase 1, our primary interest was on the experiences that participants encountered. Participants were prompted to verbalise everything they were doing and their rationale, which followed a thinking aloud strategy (Nielsen, 1993). First, users were instructed to perform the HoloLens's internal calibration function. Afterwards, the basic workflow of the HoloLens application was explained (see Fig. 7.4 for detailed workflow description), and the user immediately started the set-up. Meanwhile, two university staff members made observations and summarised them in notes (see, for example, Figure 7.5). After a short break, the participants were interviewed for about 30 minutes. The interview focused on user-experience and expertise-sharing issues. Participants also provided feedback on what was unclear in the system and offered possible improvements.

7.5.4 Data Analysis

Only interview data and observation notes were analysed in Phase 1 because the study's main purpose was to assess the user experience and the usability problems they identified from their perspective. Interviews were transcribed and submitted together with the field notes for thematic analysis, following Braun and Clarke's (2012) approach. We used an inductive coding approach and had all transcripts coded in three passes by three different researchers, who came together from time to time to discuss evolutions on the code scheme and alignment. When a code was created, a code definition was also elaborated. Finally, a joint discussion ensued for each code, and the essential codes were summarised under six themes. The citations of these codes are listed in this paper. The six themes are as follows: (1) time relief for experienced and inexperienced operators, (2) fit to work practices, (3) encouragement of independent learning of standard procedures, (4) technological constraints compensation through complementary information, (5) flexibility in the authoring process, (6) analysis and configuration of authored set-up content. We address each of these themes in turn below.

7.5.5 Limitations

Worth noting is that the findings presented in the following sections cannot be generalised. This limitation is widely acknowledged and accepted in qualitative studies (Bryman, 2008). However, we have been careful to address issues of trustworthiness and authenticity (Guba, 1981) to assure the quality of our research. To ensure such quality, we chose two strategies. First, we used different data collection methods to allow for triangulation during the data analysis (i.e. cross-checking the consistency of the findings resulting from the data collected from these different methods). Second, we drew on a systematic data analysis technique to support the generation of the findings.

7.5.6 Results

The following sub-sections introduce the findings concerning the six themes identified during our data analysis. As we elaborated on each theme, the answer to our first research question—*How can AR-based CPPS support workers in creating and consuming set-up instructions?*—was drafted.

(i) *Time Relief for Experienced and Inexperienced Operators*

During Phase 1 of our study, we received positive feedback about the potential use of our solution for KES. P14, who is a foreman in Company B, participated in the evaluation of the writing mode. He stated:

> *I can benefit from this because, if my colleagues work with the HoloLens, I don't always have to dedicate myself to [helping] them, and I can concentrate on other tasks (P14, Machine Operator, Company B, Writing Mode Test, Debriefing Interview, Phase 1).*

Experienced operators are concerned with the time they must devote to teaching novice operators the tools and tips of all set-up processes. Arguably, a motivation arises for those operators to use part of their working time to prepare the instructions. Time is also a concern for novice workers, who see the availability of a KES system as a source of relief and independence for themselves and experienced workers. P27, a less experienced machine operator, who tested the reading mode of the referred tool, explained:

> *[With the system, the] experienced colleagues can then be relieved. [In turn,] I can consume and implement the instructions [available in the system] independently (P27, Machine Operator, Company C, Reading Mode Test, Debriefing Interview, Phase 1).*

The one-time preparation of the set-up instructions consistently relieved experienced workers from the burden of teaching novice colleagues (Nilsson & Johansson, 2008). The argument is that the set-up processes can, to a large extent, be standardised, and essential instructions can be produced to support novice workers. This information corroborates findings from the ethnographic work done for the design of the referred tool.

(ii) *Fit to Work Practices*

The evaluation in a real industrial environment at different companies shows to what extent the digital AR content must fit into the working environment. On the one hand, the AR visualisation must support the work (characterised by mechanical hand movements) in the best possible way by allowing the operator to consume the set-up information in parallel to the actual activity.

> *The parallel use of the instruction to the actual activity must be learned. The operator must be able to concentrate on both the information and the activity. The simple representation of the holograms ensured that the handling of the HoloLens felt intuitive and easy to understand (P4, Production Engineer, Company A, Reading Mode Test, Debriefing Interview, Phase 1).*

Additionally, the operator's ability to perceive the peripheral environment through the visualised information must be ensured. P34, a person responsible for the operators expressed concerns in this regard:

> *Depending on where you are in the machine, the thought comes to mind: 'I hope I don't bump into anything.' This is especially dangerous with our machines and tools. There are pillars and other components that protrude into the working area (P34, Production Manager, Company D, Reading Mode Test, Debriefing Interview, Phase 1).*

(iii) *Encouraging Independent Learning of Standard Procedures*

Furthermore, the findings indicate that an added value is placed on using AR-based instructions for learning in the workplace. Participants recurrently raised the issue of standardising learning, which can potentially lead to standardising work processes. Standardisation is discussed in the literature as a key aspect of performance improvement in production (Shingo, 1985). P33 summarised:

> *The technology could guarantee the maintenance of the established standard. That all employees learn the same thing. A new standard can be easily defined with glasses and then quickly consumed. The direct target-actual comparison of the holograms with the real set-up process implementation is desirable (P33, Production Engineer, Company D, Reading Mode Test, Debriefing Interview, Phase 1).*

Additionally, the support for autonomous learning (Paas et al., 2003) seems to be much appreciated. Workers suggested that they would '...*wear the glasses as long as it takes. Until [they] have understood the step' (P32).* The constant presence of the instructions and the access to the set-up information independent of the person makes learning easier, even for people who need more time to comprehend. P13 stated:

> *This allows people who may not be able to learn as quickly to handle the set-up. So, it's very practical, and people who are a bit harder on the set-up process always have access to all set-up information and can call up the information as needed (P13, Machine Operator, Company B, Reading Mode Test, Debriefing Interview, Phase 1).*

The benefit of such instructions is not limited to the learning of set-up processes. It also helps to actively remember set-up processes that are rarely carried out. Against the backdrop of an increasing range of products and product variants (Palanisamy & Siddiqui, 2007), instructions are helpful—even for experienced set-up operators:

> *It has happened before that I had to set up a very rare product. So, I had to think and remember the set-up process. This kind of set-up support helps and gives some security (P30, Foreman, Company D, Reading Mode Test, Debriefing Interview, Phase 1).*

However, concerns emerged:

> *If I use the HoloLens and have successfully carried out the set-up process, then I have actually only followed the instructions and have not done anything myself. I then didn't have to think about anything myself, and so I don't build up any knowledge because I know that the HoloLens is available all the time (P33, Production Engineer, Company D, Reading Mode Test, Debriefing Interview, Phase 1).*

In this statement, P33 expressed doubts that using the instructions progresses learning because the user feels secure through the constant availability of the HoloLens and does not build up personal knowledge.

(iv) *Technological Constraints Compensation Through Complementary Information*

Our findings also highlight essential issues of motivating workers to adopt such a system. According to the participants, the tool should be easy to use and not disrupt the work process significantly, as the literature has widely acknowledged (Bhattacharya & Winer, 2019; Mohr et al., 2015). We found evidence that the referred aid responds well to this demand. Operators recurrently praised the fact that learning to interact with the system was not a problem. They could grow accustomed to the HoloLens Air Tap interaction without great effort, although they had never used HDM technologies before, suggesting that it would not be a barrier to accepting the technology at the workplace. Furthermore, the possibility of creating videos and holographic representations to demonstrate particular issues of specific set-up operations was highly appreciated, although findings suggest that workers would assign greater value to video material, as P9, a machine operator at Company B, who participated in the evaluation of the writing suggested in the debriefing session: *'The video provides tips and tricks. The AR presentation and the animations I have added are rather the basics.'*

Videos can capture the exact handling of the tools and, thus, convey know-how, including special information that is challenging to exchange by other media. The multimedia structure, including videos, therefore supports the organisation in building its own repository of implicit and explicit knowledge (Bloodgood & Salisbury, 2001). However, due to the current system's limitations, animated holograms have been associated with more basic knowledge. Findings suggest that the limited accuracy of the resulting holograms played a role: *'It didn't bother me so much, but the tools weren't exactly where they were supposed to be. That was a bit off' (P10, Machine Operator, Company B, Writing Mode Test, Debriefing Interview, Phase 1).* This limitation highlights the relevance of providing tools for workers to share precise information about set-up operations. A difference of 1 cm in the positioning of a part can sometimes lead to results that do not comply with the quality standards that the resulting piece must reach (see Chap. 4).

The required accuracy to successfully carry out the step was currently not in place. P31 elaborated:

> *The accuracy of the instructions is not sufficient. I've only seen about where it's going. But where exactly, I don't see. I have seen that it comes somewhere in the middle, but nothing more. I have to recognise every screw (P31, Machine Operator, Company D, Reading Mode Test, Debriefing Interview, Phase 1).*

For example, if operators look at the components to be assembled, they will possibly realise that the tool components must be mounted in a direction-oriented manner. However, the components are almost symmetrically constructed. This situation can lead to confusion and, consequently, errors. Therefore, the argument arises that holograms do not make small geometric differences visible. This finding resonates with that of Newell et al. (2014), who found that neither paper-based nor AR-based instructions could account for small details that impact task performance. It also adds to the body of research demonstrating that AR-based research does not always provide the best results in supporting assembly activities (Kosslyn, 2006; Werrlich et al., 2019). The findings raise the implication for further design and development of AR-based technology for manufacturing contexts, as discussed ahead.

However, other participants reported that even if the holograms were not precisely placed by the HoloLens, they still could determine the correct position, which is attributed to existing experiential knowledge. Our results provide an interesting insight into how the individual information components of the set-up instructions are perceived. In addition to the above quote stating that the holograms contain the basic information and do not convey all details necessary for a secure execution, P3, a machine operator at Company A, who participated in the writing mode evaluation, suggested this in the debriefing session: *'So you might want to watch the video afterwards [after watching the AR animations]. It wasn't always necessary, but I did it again just to be sure.'* Notably, the video is based on AR animations and provides additional security to conduct the set-up step correctly. However, the security thus achieved contrasts with the required speed in the execution of the set-up process. Therefore P6, a foreman from Company A, commented as follows in the debriefing session: *'I only watch the video when there are special features because it takes a lot of time.'*

Despite the shortcomings regarding the accuracy of the holograms generated with the help of the tested system, participants showed confidence in the tool's potential. P3, a foreman with more than 10 years' experience within machine set-up, said

> *One can already transfer experience. Most of the content is captured by the video. I can integrate this video recording and creation into my everyday work (P3, Foreman, Company A, Writing Mode Test, Debriefing Interview, Phase 1).*

Hence evidence shows that the approach of the referred system is predicated on the opportunities to generate different types of AR-based instructional material. For example, that videos for HMD technology can be easily associated with particular set-up operations and real-environment 3D-model animations (which are constructed with the help of the system) is promising and should be further investigated.

(v) *Flexibility in the Authoring Process*

For authoring AR holograms, the study shows that the requirements to realise authoring in a production environment is difficult to implement in their entirety.

The current tool allows the display of CAD data and the addition of linear animations of the individual tools, as well as their subassemblies (e.g. screws and washers). P30 (Foreman, Company D, Reading Mode Test, Debriefing Interview, Phase 1) emphasised that this representation is not sufficient in all set-up cases:

> *Sliding movements are simple and easy to recognise. But we have cases where several types of movement are superimposed and are crucial for correct assembly. Only if we also display these, we can enable a smooth set-up process for the novice.*

Overall, the authoring concept based on the representation of CAD data is perceived as positive, as *'[the] holograms allow geometric details of the tools to be seen' (P30)*, but what the contact point between the tool to be assembled and the already assembled component looks like is unclear:

> *The stop of the component is not recognisable. How do I know where my tool has to contact the already mounted tool? Here there are four stops; on another tool, there are only three. You have to make these stops visible. (P35, Development Manager, Company D, Writing Mode Test, Debriefing Interview, Phase 1).*

In addition, an expert's uncertainty can arise when recording the content:

> *With the video, I was no longer sure afterwards whether I had recorded the process appropriately. In the end, I can only check that if I use the instruction myself. Therefore, I have to have the possibility to adjust the content (P32, Machine Operator, Company D, Reading Mode Test, Debriefing interview, Phase 1).*

(vi) *Analysis and Configuration of the Authored Set-up Content*

The set-up configuration is considered useful because the recorded set-up data allows better analysis and—especially compared to SMED (Shingo, 1985)—a faster analysis of the set-up processes. In addition, the changed instructions can be adapted collaboratively and ad hoc transferred to the AR-Glasses. As P32 reported, the recorded content must also be checked and adjusted if necessary before it is passed on. P2 (Machine Operator, Company A, Writing Mode Test, Debriefing Interview, Phase 1) illustrated that these optimisations are absolutely useful:

The analysis of the set-up processes on the basis of paths, path times and set-up times is helpful and takes place during the analysis and subsequent optimisation of the set-up process. This is important because we are always under time pressure.

The economic necessity of continuous optimisation of the set-up process is emphasised against the backdrop of international competition and increasing product diversity (Ludwig et al., 2016; Van Goubergen & Van Landeghem, 2002). The following quote by P34 (Production Manager, Company D, Writing Mode Test, Debriefing Interview, Phase 1) shows that the different experiences and procedures of the operators are considered:

In the analysis, the procedures of colleagues can be observed, which one has not carried out oneself and has never thought about it. But they are not so bad, and you can draw a new standard from them.

The results of a long-term study, as envisioned in the third step of the DCS framework (Wulf et al., 2015), can potentially uncover other opportunities and tensions from the adoption of such technology in the workplace. Our contribution demonstrates the potential of this technology to further motivate HCI researchers in pursuing its design opportunities in industrial contexts.

7.6 Phase 2: Laboratory Test

In addition to the in-depth study at the companies on site, we also conducted a lab study to answer the second research question: How can such systems support workers with improving their task performance? This study focused on the task performance aspect and highlighted how instructions created in an authoring environment affect the task performance of novices. Our study not only investigated the possibilities of an AR visualisation but also those of the accompanying sensory verification possibilities.

7.6.1 Study Design

The second phase of the study was designed as a within-participant experiment, also predicated on cooperative evaluation. In this phase, we paid careful attention to aspects of novices' task performances with an authored AR set-up instruction in the workplace. We tried to confirm some hypotheses generated from the findings in the first phase of the study and some previous studies: for example, (Baird & Barfield, 1999; Henderson & Feiner, 2011). Overall, we tested three hypotheses:

- H1: The number of errors will be lower when workers use authored AR-based instructions during machine set-up.
- H2: The execution time will be lower when workers use authored AR-based instructions during machine set-up.
- H3: The Rating Scale of Mental Effort (RSME) value will be higher when workers use the authored AR-based instructions during machine set-up.

The number of errors and execution time are traditional indicators of user task performance, with a lower number of errors and reduced execution times associated with enhanced performance (Nielsen, 1993). Based on previous studies (Funk et al., 2016), we hypothesised that the mental effort required to work with the HoloLens would be higher due to the technological layer added.

Our independent variable was the type of support provided to workers, which had two conditions: paper-based and AR-based (authored content). The dependent variables were execution time, the number of errors and mental workload, which we measured through the RSME (Zijlstra, 1993).

Although the question may arise whether a control group personally trained by an expert would be more appropriate for a study such as the one presented, our decision for a paper-based control group was informed by findings from the literature—see, e.g. Chap. 4—and our own findings from the pre-study, which we carried out to understand the users' contexts and practices. These findings suggested that KES in companies like the ones we have addressed is mainly initiated and maintained by expert-created paper-based documents. Although new employees do receive some training from experts at the beginning of their employment, training only plays a marginal role in the KES at such companies, as it goes well beyond the limited training sessions received. It takes considerable time for new employees to develop their expertise in how to set up the assorted machines available in the company for the different types of manufactured products. From our observations, KES actually happens by reading documents scattered across the workstations and finding experts who can further clarify unclear issues in the documents. These approaches are very characteristic of KES processes, as Ackerman et al. (2013) discussed.

7.6.2 Participants

We recruited 16 participants. Again, all recruited participants were male, confirming the trend observed in the first study that men primarily carried out machine set-up. Although we do not investigate any gender-related issues, researchers may feel compelled to further investigate this trend and any issues stemming from it. Thirteen participants were students from different courses at the local university, and three were working students from local companies. Nine of the participants had prior experience with machine set-up, which ranged from one to 14 years ($M = 4.1$, $SD = 4.14$) and was obtained through previous employment in the manufacturing sector ($n = 3$) or training during their university courses ($n = 6$). Seven participants had never carried out a set-up process before. We made this deliberate decision in order to investigate how participants who lack experience with the set-up process would fare with the evaluation task of the reading mode. Students were recruited from the researchers' personal contacts, whilst the workers were recruited from the same companies that participated in the study's first phase. As mentioned before, the heterogeneous group of respondents enabled a better representation of different actors in the system (Sharp et al., 2006).

Participation was voluntary, and participants were free to decline joining in the study. They received information about the study prior to their decision to participate and signed an informed consent form, according to the basic principles of ethical research. Participants were not financially compensated for participating in the study.

7.6.3 Procedure

In Phase 2, we decided to omit the thinking aloud approach because we were interested in task performance issues. So, to minimise interaction flow disturbance, participants were not required to speak the entire time. However, we still generated fieldnotes based on observing unusual happenings that we wanted to clarify by the end of the session. Spontaneous comments from participants were also captured by a microphone attached to them and recorded.

Another difference between Phase 1 and Phase 2 was the nature of the tasks. In Phase 1, participants conducted a set-up process in their workplace as part of their productive activities. In Phase 2, the tasks on the task sheet concentrated on set-up operations—e.g. mounting or dismounting a part in the machine—and were carried out twice by all participants (once using our prototype and once using only paper-based support for the tasks). The order of the conditions (paper-based vs AR-based) was randomly assigned to counterbalance the possible learning effect stemming from repeating the task (Blandford et al., 2008). All participants performed the same tasks for the paper-based and the AR-based conditions in an attempt to increase the reliability of the results. After each task execution, the Rating Scale of Mental Effort questionnaire was administered for measuring the mental workload. After completing the last questionnaire, participants had a short break before filling out a user-satisfaction questionnaire and being interviewed.

For H1, we needed to measure the number of errors. The number of possible errors depended on the tool. Each tool had an exact assembly position and an exact assembly orientation. Any deviation from this exact position and orientation led to an error. As a result, several errors could occur with one tool. The set-up process's sensory check was deactivated to compare the paper instructions and the AR instructions. After the experiment, the functionality of the sensory check was demonstrated to the test persons and discussed in the interview.

7.6.4 Data Analysis

In Phase 2, we concentrated on the quantitative analysis of the collected data to test a series of hypotheses that we generated from the findings obtained in Phase 1. We considered the within-participant design and followed (Agresti & Finlay, 2008) the Shapiro-Wilk test to observe the normal distribution of the tested constructs. Repeated measure t-tests were used to compare mean values when normal distribution was found in the data. Additionally, we used the Wilcoxon signed-rank tests when normal distributions were not found in the data.

7.6.5 Limitations

Despite the well-elaborated study design, we are aware of our study's limitations. First, the small number of participants prevented generalising the results. Furthermore, the small sample might have affected the distribution of data available for the test construct, preventing the use of parametric tests (stronger than non-parametric tests) for testing some of our hypotheses (Agresti & Finlay, 2008). This limitation becomes evident in Sect. 7.5.2. Furthermore, using similar tasks for both test runs could have potentially influenced the results, even with the manipulations aimed to minimise the effects of confounds (Blandford et al., 2008). Although both Phase 1 and Phase 2 took place under the heading of KES, neither aimed to quantify the learning process supported by AR-based CPPS. A longitudinal study in a controlled environment is necessary for an accurate investigation of the learning process. Nevertheless, we are confident that our findings advance the state of the art by understanding how AR-based CPPS can be useful for KES in industrial contexts.

7.6.6 Results

The results of our study's second phase bring insightful findings on the impacts of AR-based instructions on task performance when created in an authoring environment. We tested a number of hypotheses, which were informed either by the literature or findings from Phase 1 of the study—sometimes by both of them. For example, widely accepted is that with the practice of a particular task, the practitioner will improve and, consequently, will be able to perform it faster and with fewer execution errors (Chandler & Sweller, 1991). The findings from Phase 1 support this claim, as seen in the quote below:

> *If you're new to the department, [having] the instructions [is] useful. But if you repeat it two or three times in a row, you don't need them anymore (P3, see earlier, Debriefing Interview, Phase 1).*

Indeed, the analysis of our quantitative data shows this tendency (see Fig. 7.6). A Shapiro-Wilk test on the data collected in the second phase suggested that the data testing H1 followed a normal distribution ($W = .954$, $p = .563$). Therefore, a repeated measures t-test was carried out. The analysis suggested no significant difference between the scores ($t = -.18$, $p > .05$, $n = 16$). A closer look at the results showed that the number of errors in the paper-based condition ($M = 2.13$, $SD = 1.71$) and the AR-based condition ($M = 2.25$, $SD = 2.18$) was indeed very close. Therefore, the null hypothesis should be accepted, consequently disconfirming H1.

The data for H2 was found to be normally distributed by a Shapiro-Wilk test ($W = .951$, $p = -504$). Thus, a repeated measures t-test was carried out, and the analysis showed no significant difference between the scores ($t = -1.60$, $p > .05$, $n = 16$). In fact, when the average time and standard deviation between the two conditions are compared, the average time in the AR condition is noticeably and slightly higher (M = 0:16:15, SD = 0:07:43) than for the paper-based condition (M = 0:13:00, SD = 0:04:23). Hence, H2 is not supported and should be rejected.

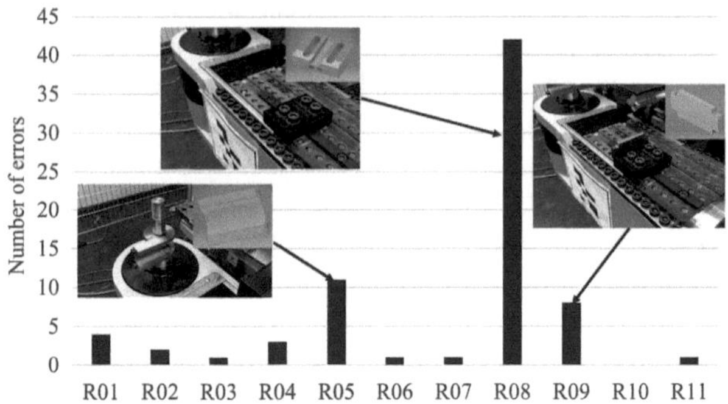

Fig. 7.6 Number of errors for each set-up step

Fig. 7.7 Performance tendency

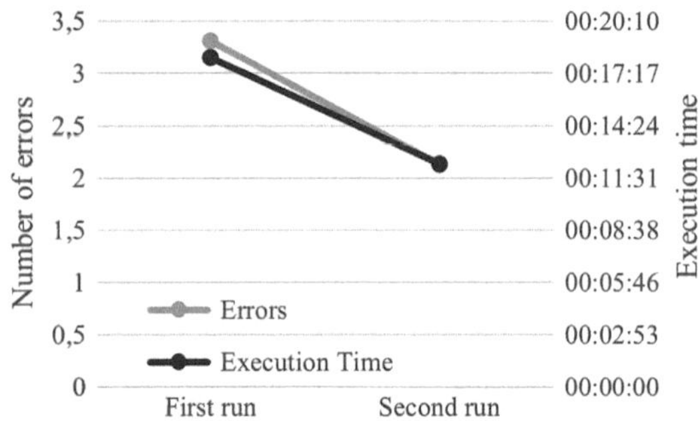

In terms of our third hypothesis, the results of a Shapiro-Wilk test provided evidence that the data available to test H3 did not follow a normal distribution ($W = .780$, $p < .001$). Therefore, a Wilcoxon signed-rank test was performed. The analysis of the results indicated that the mental effort in carrying out a set-up process using AR-based instructions was indeed considerably higher ($M = 31.25$, $SD = 21.72$) than the effort required in the paper-based condition ($M = 23.44$, $SD = 18.77$, $Z = -2.41$, $p < .05$). Therefore, H3 is supported by the findings.

In summary, our results suggest that the referred authored AR-based instructions do not promote any reduction of the number of errors compared to the paper-based instruction and that the set-up processes carried out by following that type of instruction demand more time and a higher cognitive load. This finding can be potentially associated with two issues. First and foremost, the myriad of materials available in the AR-based approach may well overload the users (Funk et al., 2016). After all, the system offers users numerous possibilities: for example, watching a video, reading textual instructions and observing an operation through a 3D-overlaid hologram. These conditions may require more attention and cognitive processing compared to simpler and more objective paper-based instructions. Second, the system's limitation concerning accuracy (identified in the first phase of the study) potentially skews the results. Furthermore, the authoring approach presented here seems to prevent sufficiently detailed instruction. Although 3D models of the tools are used, the representation is insufficient, especially for very symmetrical tools (see Fig. 7.7). Therefore, additional information is necessary. Videos convey essential set-up information, especially in the case of complex assembly movements that could not be created with the help of the authoring environment. However, videos still create a higher cognitive load.

The use of sensory monitoring is particularly motivated by the final securing of the tools' set-up positions on the machine and the tool size. Thus, it allows for verifying human execution of the set-up process guided by the set-up support. Already in Phase 1 of the evaluation, P30, a foreman in Company C, emphasised the following: *'The security aspect is important. It must be ensured that the operator does not forget anything. Otherwise, serious damage can occur.'* Particularly against the backdrop of accepting instructions, a sensory recording of correctly executed set-up operations is helpful. Otherwise, errors are attributed to insufficiently precise instructions. Consequently, the acceptance of the instructions suffers enormously. The HoloLens's tracking technology does not yet achieve the necessary accuracy to ensure the described checks (Evans et al., 2017), which users see as a necessity:

> *The camera checks after the set-up process must ensure that the manual set-up operations have been carried out successfully. Otherwise, the instructions are not sufficiently precise (P43, Debriefing Interview, Phase 2).*

The results of the sensory testing should be integrated into the AR interface. Even if external sensors are involved, an interface to the HoloLens is desirable. P45 (Debriefing Interview, Phase 2) hinted: *'You should strive for a single interface. Everything must be visible on the HoloLens; every other interface is a distraction.'* Given the findings regarding the RSME, the number of interfaces must be reduced in any case.

Furthermore, we assert that for novice workers, one of the most vital aspects concerning KES is easy access to trustworthy learning material. Knowing that the content has been generated by an experienced colleague vouches for the technology, and it can, therefore, support its adoption in the workplace:

> *It's probably easier to consult the glasses. You don't have to bother a colleague. I trust my colleagues and the glasses equally as long as I know that the content on the glasses has been created by my colleague (P13, Machine Operator, Company B, Reading Mode Test, Debriefing Interview, Phase 1).*

Therefore, the authoring environment and the direct use of the expert-created instruction ultimately leads to the novices' higher acceptance of the instruction.

7.7 Discussion

As initially argued in the paper, machine set-up is significantly relevant to time-critical processes in manufacturing contexts dealing with the production of assorted products. The past few decades have witnessed an increased number of workplaces that, in response to dynamic markets, have invested in producing a broad array of articles—many times produced on the same machines. In order to produce these articles, machines must be set up both physically and digitally, requiring assorted types of knowledge and expertise to deal with the ecology of articles, tools, machines and practices that workers interact with as they engage in their productive activities. The high rotativity of workers and the complexity of the workplace ecology demands effective ways to share knowledge and expertise.

This paper has addressed issues of technological mediation in KES concerning machine set-up by looking more closely at creating and consuming the instructions. These factors are a vital part of technology-supported KES. Existing studies on AR-based support in manufacturing have usually focused on laboratory experiments, neglecting potential impacts of the natural setting in the use and performance of the technology (Bhattacharya & Winer, 2019). Our study has combined both a qualitative and quantitative approach for investigating issues of authoring and reading instructions in manufacturing supported by AR-based CPPS technology. As evident in the results, relevant issues emerged concerning the authoring of instructional content in manufacturing contexts with tools (as the ones designed and investigated in our study), which should be considered for designing and developing new and innovative tools enabling KES support for workers of this sector.

7.7.1 Trade-offs of Producing and Using Content for AR-based CPPS

In trying to answer our first research question—*how can an AR-based CPPS, as the one explored in our study, support workers in creating and consuming set-up instructions?*—we have been confronted with a series of trade-offs concerning the production and use of AR-based content in manufacturing contexts for CPPS.

The practical and industry-oriented evaluations of the authoring component of the referred AR-based CPPS have shown that workers are strongly motivated to create instructions and use available documentation on the shop floor. Our results have demonstrated that investing time to create documentation would mean reducing interruptions in the future. Consequently, workers would be better able to concentrate and perform personal work tasks in conformance with previous findings from the literature (Speier et al., 1997). Despite the focus that companies have given to paper instructions (which are arguably cheaper to print but not necessarily cheaper to create), this type of support proves to be cumbersome in industrial settings, both during creation and handling. Similar results have already been discussed in the literature but not in contexts presented in this study (see Chap. 4). Therefore, this finding, although not necessarily generalisable, can be transferred to other contexts that share similarities with the one that has been explored.

In the following section, we further elaborate on the answer to our first research question. We first focus on authoring instructions with the help of the AR tool support that we have designed and then on the use of such content.

7.7.2 Authoring Set-up Instructions

Many issues arise in creating and authoring instructional content to teach colleagues the tricks of the trade, which range from the tools to be used to the way that the content will be structured and presented—see, e.g. (Funk et al., 2016). Ultimately, these uncertainties lead to strongly fluctuating execution times and numerous errors associated with task accomplishment. Hence, these consequences uncover a need for tools to support workers with practices concerning KES (see Chap. 4). The results of the study show that for the different information levels involved in KES, three essential contents must be generated with the help of the authoring environment.

First, accurate 3D data should be used (see Chap. 8) in conjunction with animations to illustrate the exact motion sequences. Therefore, an authoring environment for set-up processes should focus on creating accurate animations. The animations turned out to be a critical element because they approximate the virtual representation to the real activity of the set-up and enable a direct comparison of the activity with the virtual representation. This result resonates with existing findings in the literature (Newell et al., 2014; Rzayev et al., 2018). Indeed, existing literature has focused on the creation of 3D content rather than on the creation of appropriate animations (MacIntyre et al., 2005; Nebeling et al., 2018; Nebeling & Madier, 2019). In tool-based set-up processes, the 3D data is often available and, therefore, does not have to be created. Instead, the data is animated and put into a sequence.

Second, additional information in the form of videos should be used if detailed information is not included in the virtual 3D data. Videos effectively complement the information presented in the holograms. The evaluation shows that the holograms provide the basics, and the videos provide the exact handling of the individual components. In addition, videos can be recorded quickly and easily with the existing hardware. However, videos have limitations, as the set-up process's performance suffers.

Third, the evaluation shows that in addition to the tool to be assembled, its support surfaces and attachment points should also be highlighted.

Although we arrived at these findings from the evaluation of the tool support, they can be considered implications for design for other KES to be used in similar contexts as the one explored in our study. Put differently, in order to better support KES in contexts highly associated with embodied knowledge (for example, the knowledge embedded in how to turn parts or fix parts together), one should think of using AR-based technology aggregating 3D data, videos, animation, and digital twins. An example would be the design of tools to support construction-site workers, who commonly assemble elements as they proceed to build something.

Worth noting is that despite the fact that authoring becomes a bit more complex due to the creation of these three essential components, the generated content benefits from being expertly created and, thus, a similar trustful learning environment emerges, as seen in personally guided learning situations (Büttner et al., 2020).

In addition to the specific findings regarding AR aspects of our tool, general implications can also be derived. The results of our study show that, independent of being provided through AR technology or not, set-up instructions must be introduced in a way that minimises disruption to workflow both on the part of the experts and the novices (Speier et al., 1997). In order to achieve this goal and create an autonomous learning environment, instructions should correspond to the levels of detail listed in the evaluation (Bresciani & Eppler, 2009; Paas et al., 2003) and could be adapted for both experts and novices (Kalyuga et al., 2003). In this context, the role of the technology would be to provide an environment for the rapid creation of instructions by experienced people (Nebeling & Speicher, 2018). Therefore, existing practices may be conveyed to use the value of the existing resources (Bloodgood & Salisbury, 2001). However, our findings highlight that these tools must be trustworthy, as evident in the results presented throughout this paper. This trust can be obtained through different levels of information made available by the system, in a way to communicate the basics and the details of the set-up process, reinforcing previous findings from the literature (Korn et al., 2013).

7.7.3 Using Authored Set-up Instructions

Our findings suggest that content generated once in an authoring environment does not adequately represent the knowledge recorded by experts. Although the instructions are generally trusted, room for improvement is evident when analysing the generated content. Based on the analysis, the content must then be adapted. Process data turns out to be beneficial for the analysis. The data used here in the form of set-up times and set-up paths must also be recorded in the authoring environment and subsequently combined with set-up experts to form an optimised set-up standard. Bhattacharya and Winer (2019) used this strategy, which addressed some of the user needs we identified. According to the empirical results, the definition of a generally valid and almost optimal standard with the help of the CPPS set-up support presented in this paper is rather beneficial to KES. However, the possibility of simple representation of set-up standards is also seen critically: namely, precisely when the learning process of novices is prevented. P33 pointed out that consumption does not necessarily contribute to knowledge building. The constant reliance on technology can, in fact, work against learning. Accordingly, it can prevent the novice from gaining the embodied knowledge concerning the set-up process, which needs to take place for effective learning (see Chap. 4). Therefore, our findings suggest that controlled use of the technology must take place. Furthermore, a usage plan or a training strategy must be implemented. The user must be motivated to see the technology only as a temporary solution. Other researchers and practitioners can use this finding in the elaboration of technology to support KES: these technologies must support the user to appropriate the shared knowledge and find innovative uses for it. Likewise, users should not depend on its use.

In addition to the learning incentives, the machine operator must receive prompt visual feedback. From our observations, a direct comparison should be made possible by the superimposed AR visualisation. This approach would facilitate the understanding of instructions, even though workers have mentioned that they would associate the visualisations to more basic information. As we argued before, if the accuracy is insufficient, workers will tend to de-value the visualisations. As a measure to mitigate this, additional sensory monitoring options can be installed to provide prompt feedback on all activities performed (Goomas, 2012; Hermsen et al., 2016), enabling more effective immediate feedback (Goomas et al., 2011). In addition, the evaluation results show that the better the holograms embed themselves into the machine's environment, the better the instructions can be consumed. In general, the holograms based on the CAD data are perceived as very detailed, as all-important geometric information is included. Alternatively, due to a high level of immersion, the perception of the periphery risks being reduced, which is particularly the case with very nested machines: see, for example, Figure 7.5 (Gonzalez et al., 2019). In such cases, user safety must be considered (for example, sensory monitoring of the workspace).

In general, we assert that the addition of sensory monitory options follows the direction of a CPPS approach to the design of KES systems. Such approaches have been found to be a key aspect of what has been termed the fourth industrial revolution (see Chap. 4). The results of the study align with the findings of the OM literature, which has stated that formal approaches and simple data storage lack promise (Randall et al., 1996). According to our study, visualisation technologies need to be linked with sensor technologies. Through these measures, the operators are supported in the cognitive work, and their existing abilities are augmented through the use of a socio-technical support system (Cimini et al., 2020). Looking at the results against the backdrop of a KES process shows that socio-technical support systems do not primarily offer support through the visualisation of standard procedures but enable inexperienced operators to make their own experiences based on the instructions. Thus, operators build up their own knowledge base.

However, worth noting is that the role of prior knowledge cannot be precisely defined based on our study, but the results show that different information densities are necessary for different users. Whether the information density correlates with prior knowledge cannot be determined from this study. Understanding the full potential of such technologies and the changes they are about to bring uncovers an urgent need. Knowledge transfer models like the one introduced in Chap. 5 may be very helpful, although they still require scrutiny.

7.7.4 Performance Enhancement is not a Given

In pursuing the answer to our second research question—how can such systems support workers with improving their task performance?—our findings demonstrate that simply using the type of AR-based CPPS that we have explored in our study does not guarantee enhanced performance. This finding aligns with findings from past and current studies: for example, (Kosslyn, 2006) and (Werrlich et al., 2019). Furthermore, we have observed that the notion of enhanced performance is quite complex.

Our quantitative findings suggest that the sort of AR-based instructions investigated are not directly responsible for short-term learning effects, but our qualitative findings suggest that they may enable workers to carry out set-up processes effectively. In theory, a significant learning effect can be achieved when set-up procedures are repeated over and over again. Indeed, this idea is what practice, in the sense of learning, is about: mastering processes and procedures by repetition (Schatzki, 2001). Our findings show that AR technology may impact this process. A difference in the execution time between the HoloLens and the paper instruction grew clear. Evidence, albeit not statistically significant, indicated that those test persons with AR-based instructions need more time for a set-up step than without the HoloLens. In this respect, Speier et al. introduced different results that suggested the advantages of spatial representations (Speier et al., 2003). The Rating Scale of Mental Effort value is also higher, indicating a higher cognitive load when wearing the HoloLens. These results contrast with some results in the literature. Tang et al.'s study posed that AR makes it possible to reduce the cognitive load because attention no longer requires alternating between the work environment and the instruction. This change can lead to enhanced performance (Tang et al., 2003). These apparently contrasting results may be caused by differently complex tasks. The study presented in this work shows that the AR set-up instructions do not lead to improved performance, at least initially. Rather, the multimodal visualisation of the glasses causes a high cognitive load (Chen et al., 2009; Feinberg & Murphy, 2000). Both results in the context of the study may be attributed to a lack of the holograms' accuracy, as some of our participants mentioned.

Bresciani pointed out that, in such a case, the effect of an excessive extraneous cognitive load appears. This result is identified in the study by an increased Rating Scale of Mental Effort value (Bresciani & Eppler, 2009). Such results yield relevant implications that must be tackled in the referred system we tested, as well as in any system that aims to provide similar support to workers. They also raise the question of whether, with practice, using AR-based instructions would have a different effect on performance. AR instructions show a significant added value when combined with sensory values. The study shows that AR instructions in combination with sensory process values lead to optimisation and standardisation of set-up processes. This, in turn, inevitably leads to an improved set-up performance in the long term. In addition, the study suggests that a sensory check of the set-up activity (for example, with a camera) can cognitively relieve the machine operators by reducing the fear of incorrectly executing the visualised set-up step.

Another issue that future research should address concerns the extent to which the referred aid can support experienced workers to better carry out their activities. As Kalyuga et al. demonstrated, expertise reversal effects may occur in such situations (Kalyuga et al., 2003). Novices usually find instructions useful because they typically do not have a schema for coping with a given situation. Experts, however, usually have relevant schemata from their experiential knowledge. If they use instructional guidance, they may experience redundant information and, consequentially, increased cognitive load.

Finally, as already mentioned above, technology-dependence issues should also be further investigated. Past research has shown that novices who are unable to determine the set-up process on their own grow more dependent on AR technology and more intensively than do experts (Shu et al., 2011). Therefore, novices are hindered in the development of autonomous process sequences more than experts. However, novice workers profit by standards introduced with the system by often activating the schemata they acquired. This aspect raises issues concerning the extent to which this technology can really support learning and professionalisation and should be further investigated.

7.8 Conclusion

AR instructions to support assembly processes have been extensively explored in the literature with no uniform results regarding the support AR technology provides (Blattgerste et al., 2017; Büttner et al., 2016; Funk et al., 2016; Werrlich et al., 2019). The same applies to AR authoring systems. Studies indeed have shown how AR content can be generated in different ways (MacIntyre et al., 2005; Nebeling et al., 2018; Nebeling & Speicher, 2018); however, remaining open is (1) how useful the generated content is; (2) how the authoring environment can be embedded in a real production environment; and (3) how exactly the generated holograms should look. Ultimately, the use of AR technology aims to support KES processes. For this, the authoring environment and the content's playback must mesh smoothly. This need is where CPPS technologies come into play. Here, instructions are enriched with sensory values to create a more realistic image (Ludwig et al., 2017). To the best of our knowledge, no studies have combined all these fragmented elements into a comprehensive study. This gap in the literature is precisely what this paper aimed to mitigate, focusing on industrial set-up processes. Our results can potentially be transferred to other assembly processes because an essential part of the set-up includes assembly and disassembly processes. Therefore, this paper encompasses a study that evaluates the following: the authoring of AR content, the playback of AR content and the possibilities of optimising the content via a CPPS infrastructure in a real production environment.

As a summary of our results, we have elaborated a data flow of the most important data for assembly processes supported by an AR-based CPPS based on the elements introduced in Fig. 7.4. This figure lists the data objects that have found their way into the evaluation studies. The results of the evaluations show in which process steps the data was helpful and which data must be considered in the future (see Fig. 7.8).

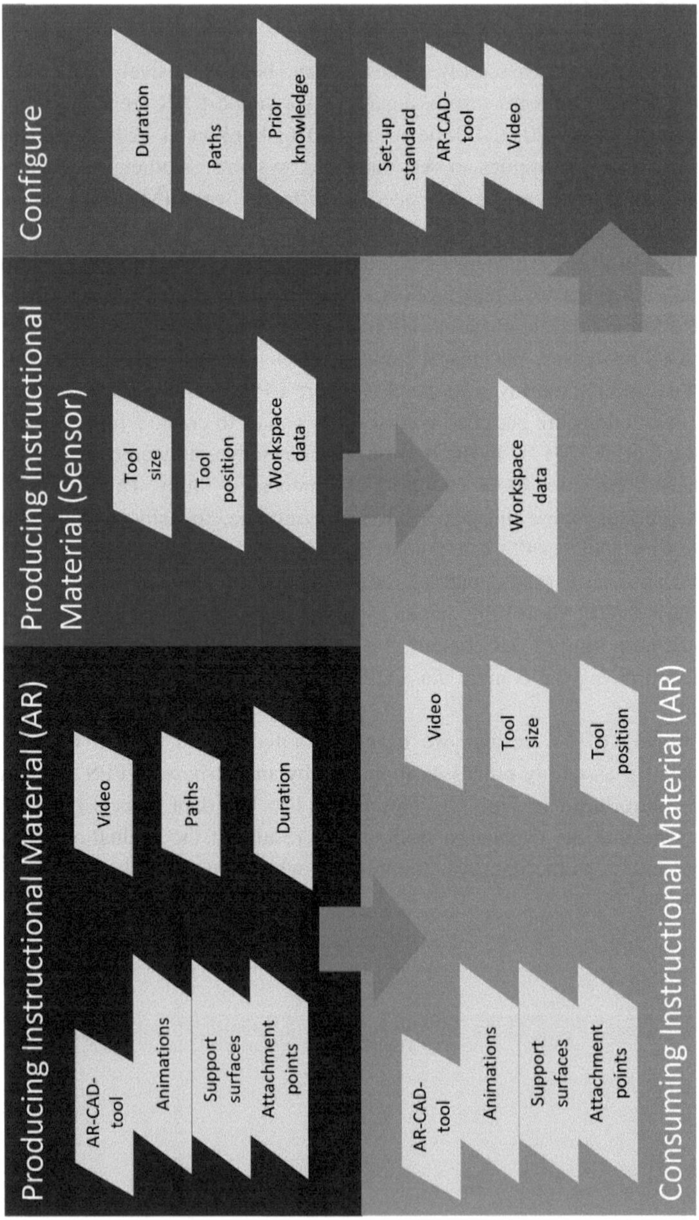

Fig. 7.8 Representation of an abstract data flow for producing, consuming and configuring instructional material using an AR-based CPPS

Our results suggest that, when producing instructional material, the most accurate possible animation of the assembly movements should be generated based on CAD data of the tools. CAD data is recommended, as more abstract geometries contain less detail, making it more difficult to compare real geometry with holograms. Showing only the geometry to be assembled is insufficient: support surfaces and attachment points must also be highlighted. In any case, videos should be recorded to prevent possible inaccurate representations of the holograms and to support detailed movements with video instruction. In addition, metadata of the assembly process, such as paths and durations, should be recorded, as they can be used as a basis for later optimisation. Sensory data should be created to support this. A check of the tool size and the tool position has been shown to be helpful. These data serve to give the user more certainty that the correct steps have been taken. The evaluation also found that the working workspace should also be sensed so that the user of AR glasses can be warned directly through the glasses if the user risks colliding with a tool or machine. Both the holograms and the sensory data should be produced in such a way that they are directly suitable for consumption. In addition, configuring the instruction based on this data should be possible. Ultimately, an optimised standard procedure should be prepared; novices can achieve the greatest learning success based on an optimised standard. However, the standard must also leave room for the operator's own experience. Sensory feedback technologies must ensure that a correct result is nevertheless achieved at the end of the set-up process. Only the users' prior knowledge of these instructions must be considered by adapting the amount of information.

Despite the limitations and the need for further research, our study offers a novel evaluation of an AR-based CPPS for the production, consumption and configuration of instructional materials and provides concrete guidance on what data should be collected and used. In the future, we aim to learn more about the exact mechanism of KES supported by the presented infrastructure with the help of a longer-term evaluation in a production environment. We expect that findings from appropriation research will further influence the use and also the benefits of such technologies (Ludwig et al., 2019). However, phenomena of appropriation can only be observed with the help of long-term studies.

8 Comparison of Different Types of Augmented Reality Visualizations for Instructions (CHI 2021)

Augmented Reality (AR) is increasingly being used for providing guidance and supporting troubleshooting in industrial settings. While the general application of AR has been shown to provide clear benefits regarding physical tasks, it is important to understand how different visualization types influence user's performance during the execution of the tasks. Previous studies evaluating AR and user's performance compared different media types or types of AR hardware as opposed to different types of visualization for the same hardware type. This paper provides details of our comparative study in which we identified the influence of visualization types on the performance of complex machine set-up processes. Although our results show clear advantages to using concrete rather than abstract visualizations, we also find abstract visualizations coupled with videos leads to similar user performance as with concrete visualizations (Fig. 8.1).

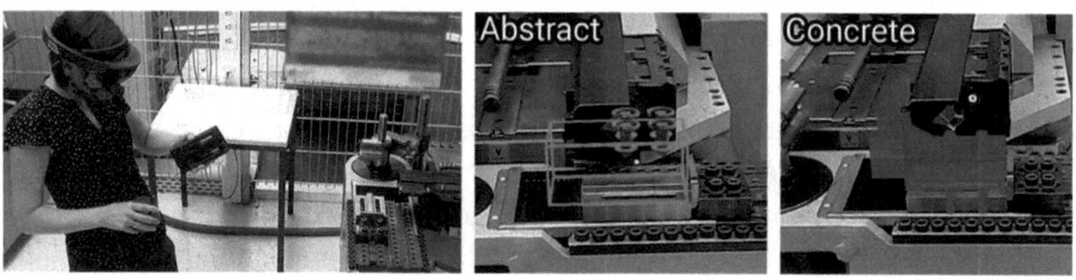

Fig. 8.1 In our study we compare abstract and concrete AR visualizations for instructions using head mounted displays

S. Hoffmann, *Knowledge Sharing on the Shopfloor*,
https://doi.org/10.1007/978-3-658-50133-4_8

8.1 Introduction

In our daily work we often find ourselves in situations in which we encounter problems. If we are unable to solve the problems ourselves, we may seek support from people with more experience and expertise (Crabtree et al., 2006; O'Neill et al., 2005). This is particularly true within complex work environments and when performing knowledge-intensive tasks. These tasks encompass non-trivial activities that have a strong contextual reference characterized by a multitude of environmental factors, requiring a significant amount of knowledge to achieve a successful outcome. Examples of knowledge-intensive tasks are setting up or retooling a modern production machine, repairing a car engine, or solving 3D printer errors.

In recent years, the use of Augmented Reality (AR)-based technologies to support remote troubleshooting (Aschenbrenner et al., 2018; Masoni et al., 2017; Microsoft, 2020) as well as digitally enriched instructions (Bottani & Vignali, 2019; Microsoft, 2020) has increased significantly for these types of tasks. Moreover, the COVID-19 pandemic and the resulting lock-down in many regions of the world have created increased interest in AR for supporting remote maintenance or troubleshooting and detailed instructions (Viljakainen, 2020). The primary advantage of AR is that it creates a 3D-based overlay of digital content onto the physical environment, which offers a rich set of additional information (Akçayır & Akçayır, 2017; Fernández del Amo et al., 2018; Jetter et al., 2018). While the general application of AR concepts shows clear benefits for supporting complex settings, its actual design and technology as well as its impacts on users are often not well established. Regardless of the chosen hardware and its advantages in specific applications (such as head-mounted displays [HMDs] for hands-free applications (Nakanishi et al., 2010; Reuter et al., 2019), handheld displays [HHDs] like smartphones for cost-effective, easy-to-use cases (Arth et al., 2015; Sanna et al., 2015), or projector technologies for longer, static, and group settings (Mengoni et al., 2018; Rodriguez et al., 2015)), a crucial factor is the actual visualization and presentation of AR content in relation to the complex tasks and knowledge-intensive work.

Existing studies evaluating different AR-based content visualizations and their performance on working practices have compared different media types (e.g., paper-based instructions versus AR-based visualization) or different types of AR hardware (e.g., hand-held devices versus head-mounted displays versus projection-based AR). Our study complements this set of existing studies by evaluating different forms of visualization on the same hardware type for performing a complex task. In particular, we examine the task completion time, error rate, and task and cognitive load of setting up a complex machine with abstract and concrete visualization types on the head-mounted display Microsoft HoloLens. The visualization types consist of abstract and rather simple holograms like wire frame boxes and 3D arrows, as well as concrete detailed 3D models from CAD data.

In the following sections, we present the current state of the art regarding the use of AR in manual assembly and maintenance tasks (Sect. 8.2). We then identify and discuss the research gap regarding the comparison of different visualization types on the same task and same hardware (Sect. 8.3). In Sect. 8.4, we describe the applications we compared within our evaluation study. We then present our study design and procedure in Sect. 8.5 and our results in Sect. 8.6. Finally, we conclude with a discussion of the results with regard to visualization types of AR applications and their implications for AR authoring tools (Sect. 8.7).

8.2 Related Work: Augmented Reality for Supporting Complex Tasks

Augmented Reality has a wide application field, particularly in the industrial context, where there is a huge interest in using AR for providing assistance for complex tasks such as manual assembly or maintenance work (Fernández del Amo et al., 2018; Wang et al., 2016). In these applications, virtual computer-generated visual features are superimposed on the real environment support the execution of these complex tasks. These AR applications commonly support workers in sequential tasks by providing context- and task-related information in-situ. Over the last two decades there have been several studies investigating the suitability and effectiveness of AR applications for manual assembly or maintenance tasks. In these studies, AR applications were usually compared to paper-based instructions by measuring the time required to complete a task, the amount of errors that occurred during the task, and the mental workload during the execution.

Funk et al. (2016) compared four different hardware types (paper, a smart glass, a smartphone, and in-situ projection) using a standardized Lego Duplo task (Funk et al., 2015) with 16 participants (with-in subject). In this study, all hardware types used the same pictorial instructions. The authors measured the task completion time, errors, and the task load. The results showed that there is no significant difference between in-situ projections and paper instructions regarding the completion time. However, the same pictorial instruction on a HMD and HHD resulted in a significantly slower time to recognize the correct elements compared to the paper. In addition, more errors were produced when using the HMD compared to the hand-held device and in-situ projection. In regard to the cognitive load, the in-situ projection was perceived to be the lowest and the HMD was the highest.

Blattgerste et al. (2018) investigated different in-situ instructions for assembly tasks by comparing four visualization types (3D in-situ, 2D in-situ, 3D wire, and side by side) using the Microsoft HoloLens in a with-in subject study consisting of 24 participants asked to solve a standardized LEGO Duplo task (Funk et al., 2015). While the results show a faster completion time and less errors when using 3D in-situ visualizations, there was no significant difference between the systems in terms of the task load. Blattgerste et al. (2017) further compared different types of AR hardware (Microsoft HoloLens, Epson Moverio BT-200, and smartphone) to paper-based instructions. The paper-based instructions resulted in the fastest overall task completion time, while the Microsoft HoloLens led to the least number of errors, but with a significantly higher cognitive load.

Smith et al. (2020) analysed the effect of different interaction modalities (touch and voice) and visualization modes (3D model, text annotation, and in-situ video) of an AR application on task load, task completion time, and error rate for 24 participants performing a LEGO Duplo task. The in-situ video led to the fastest task completion time, followed by superimposed 3D models and text annotations. In addition, the task load was the lowest for the in-situ videos and the highest for the text annotations. The method by which the users interact with the AR content (touch versus voice) had little to no effect on the measured performance (Table 8.1).

Table 8.1 Overview of the closely related studies. Most AR vs. AR studies also used paper instructions as baseline

	Study	Medium	Task	AR-Visualization
Paper vs. AR	Tang et al. 2003	Paper vs. Sony Glasstron LDI-100B	Lego Duplo	3D models and arrows
	Syberfeldt et al. 2015	Paper vs. AR-enabled Oculus Rift	3D puzzle	Abstract highlighting
	Fiorentino et al. 2014	Paper vs. Monitor-based AR	Engine maintenance	3D animations, text, voice
	Hou et al. 2015	Paper vs. Monitor-based AR	Pipe system assembly	3D models of pipes
	Uva et al. 2018	Paper vs. In-situ Projection	Engine maintenance	Test and highlighting
AR vs. AR	Funk et al. 2016	Tablet (no AR) vs. Epson Moverio BT-200 vs. In-Situ Projection	Lego Duplo	Same pictorial instructions
	Blattgerste et al. 2017	Epson Moverio BT-200 vs. Microsoft HoloLens vs. Smartphone	Lego Duplo	Same, but slightly adjusted for the different media
	Smith et al. 2020	Smartphone	Lego Duplo	3D models vs. text-annotation vs. in-situ video
	Radkowski et al. 2015	Monitor-based AR	Engine maintenance	Abstract vs. concrete
	Blattgerste et al. 2018	Microsoft HoloLens	Lego Duplo	3D in-situ vs. 2D in-situ vs. 3D wire vs. side-by-side

Radkowski et al. (2015) investigated the effect of different visual features (abstract versus concrete) of 3D content for different degrees of difficulty in manual assembly tasks. Abstract refers to visual forms (e.g., lines, colors, shapes) that are separate from any concrete 3D models. The authors conducted a between-subject study with 33 participants (abstract visualization versus concrete visualization versus paper-based). The participants had to assemble a mechanical axial piston engine with a total of 16 manual assembly process steps. Two of these steps were rated with a high degree of difficulty; all the others were rated with a low degree. In comparison to the LEGO Duplo tasks (Funk et al., 2015), this assembly task can be considered to be more difficult as it is based on real conditions. The authors used a tabletop AR workstation displaying an AR view of the working space. They showed that the abstract visualization led to an overall longer completion time and more errors compared to concrete visualizations. Further, the concrete AR visualization resulted on average in a similar completion time compared to the paper-based instructions, while the completion time for abstract visualization was almost twice as long. This result suggests that concrete visualization is more suitable for relatively simple tasks. However, which kind of visualization type is more appropriate for rather complex tasks could not be answered as there were ambiguous results regarding the time of the two difficult assembly steps.

Wiedenmaier et al. (2003) argues that users only benefit from AR when focusing on difficult assembly tasks and most of the previous studies (Blattgerste et al., 2017; Funk et al., 2016; Radkowski et al., 2015; Smith et al., 2020) confirmed that AR does not provide an advantage for the completion time of simple tasks. Furthermore, Wiedenmaier claimed that *"[g]ood manuals often use line drawings instead of photos, to reduce complexity. In the same way, assembly tasks can often be simplified by more abstract objects"* (Wiedenmaier et al., 2003). Haller (2004) concluded that 3D models look nice for AR developers, however the actual end users often ask for abstract but familiar schemes or 2D drawings. These statements contradict the results of Blattgerste et al. (2018) and Radkowski et al. (2015).

Interactivity is considered as an important factor when focusing on AR applications (Azuma, 1997), but only few studies (e.g. (Smith et al., 2020)) have investigated the impact of different interaction techniques in this field. In most existing studies regarding types of instruction systems, there is no active interaction with the system itself. Usually, techniques such as Wizard of Oz or automatic detection of the correct placement are used and the user should fully concentrate on the task. However, an automatic detection of the correct placement can not be implemented for every task, wherefore multi-modal interactions with hand gestures, voice or special controllers are meant to enhance the user experience in AR applications (Chen et al., 2017; Heidemann et al., 2004). While designing the interaction, the environment as well as the task needs to be considered. Due to loud noise, for example on the shop floor, voice interaction is often not suitable for interacting with the AR system.

8.3 Research Gap and Approach

The literature analysis reveals many existing studies that evaluate AR-based assistance for manual assembly and maintenance tasks, as shown in Table 8.1. Usually these studies either compare AR-based instructions to those using different media such as paper-based and mentor-based instructions (Werrlich et al., 2018), or they focus on comparing AR-based instructions through different hardware types such as projection-based AR, hand-held devices, or HMDs. Only a few studies compare different visualization types for instructions using the same AR hardware type.

Although the success of AR-based applications for assembly depends on many different factors, current studies almost do not adequately address the complexity and setting of the task (Dourish, 2004) and instead focus on simplified Lego assembly tasks, neglecting a variety of conditions of real complex tasks. Only one study (Radkowski et al., 2015) compares visualizations of one type of hardware in real tasks, but the static, monitor-based apparatus of the study setup and the resulting limited interaction and visualization area neglect the complexity of mobile settings as well as limit the possibilities of 3D representation.

Through our study, we contribute to the field of comparative studies of AR for assembly tasks by examining the visualization types for instructions that are suitable for HMDs in real-world tasks. To accomplish this, we conducted a user study to identify the influence of different abstract and concrete visualization types on the key performance indicators in machine set-up processes.

For our different forms of visualization, we distinguish between the types of 3D holograms. We use very simple (abstract) holograms like wire mesh boxes and simple 3D arrows as well as detailed (concrete) 3D models from CAD data; we refer to the former as abstract AR (AAR) and the latter as concrete AR (CAR). In addition, we also distinguish between the use of multimedia data through videos. Such instructions with additional video files we refer to as abstract AR with videos (AAR+V) and concrete AR with videos (CAR+V). Additionally, we used paper-based instructions as a base line. The key performance indicators of our study were the task completion time, error rate, as well as task and cognitive load.

We aim to investigate the following hypotheses:

- H1: The task completion times for the set-up procedures with AAR, CAR, AAR+V, CAR+V and paper-based instructions are significantly different.
- H2: The number of errors for the set-up procedures with AAR, CAR, AAR+V, CAR+V and paper-based instructions are significantly different.

- H3: The cognitive load for the set-up procedures with AAR, CAR, AAR+V, CAR+V and paper-based instructions is significantly different.

To verify our hypotheses, we carried out a between-subject study as the execution of the instruction with either visualization type may lead to learning effects which could influence our dependent variables. We collected qualitative as well as quantitative data sets. With regard to the quantitative data set, we defined dependent variables such as task completion time, error rate, mental workload, perceived usability and user experience and the visualization type with five characteristics as independent variable.

8.4 AR Applications

We evaluated two self-developed Microsoft HoloLens applications that support the same task of performing hardware-centred activities—but with different visualizations. The main concept of both applications is to guide users in performing a task through AR instructions. The applications were meant to support novice users in procedural machine set-up tasks.

When we compare concrete (see Fig. 8.2) with abstract (see Fig. 8.3) types of visualization, it is clear that concrete visualization is much more detailed and closer to the actual application. While in the abstract visualization only arrows or generic forms indicate points of action, the concrete visualization maps exactly the 3D models with the real work pieces, giving the user richer indications of the parts to be used and their locations.

8.4.1 Concrete Augmented Reality Visualizations

The first application (concrete augmented reality, CAR) was created in the research project Cyberrüsten 4.0 and focuses on the procedure of setting up complex machines (see Chap. 5). The application uses 3D models provided by CAD machine data and the instructions are typically created by professionals on the shop floor (domain experts) with the same AR application in edit mode. The instructions are organized in small sub-tasks as sequential step-by-step commands.

When creating an instruction with the CAR application the expert defines which component needs attention by selecting the component's name from a list, selecting a virtual overlay (shadow) of the component on the machine, then defining the mounting direction for each step. The expert can take photos or videos to further visually illustrate the activities for each step. Photos and videos are recorded with the built-in RGB camera of the Microsoft HoloLens to create the same perspective as that of the expert.

To view the instructions, a main panel shows the information about the current process step including the current step number and the name of the component. On the left side of the main panel is another panel showing pictures and videos of the current step. By default, the panels are connected to each other and follow the users' movement, but they can be fixed to any position within the room by the user. Buttons are located in the lower corners of the main panel to navigate through the instructions. Furthermore, concrete 3D models show the component and its mounting position, as shown in Fig. 8.2. The 3D models move to the mounting position to show the mounting direction, the models dwell for three seconds on the mounting position, and then the animation repeats.

Fig. 8.2 Concrete AR visualization of assembly step a) 1, b) 7, c) 11

8.4.2 Abstract Augmented Reality Visualizations

The first field trial at a local company of our first application showed that creating the AR content and preparing the CAD data required an enormous amount of effort, which may not be possible by actual domain experts. Furthermore, the company in which we evaluated the first application wanted to use it for other tasks (beyond setting up a machine) for which no CAD data were available. We therefore implemented a second AR application (abstract augmented reality, AAR) from scratch with less detailed 3D models compared to the first application which also simplified the content creation. This ensured that the actual

AR content creation could be done by non-AR experts. While building this second application, we replaced the detailed CAD models with simple abstract 3D models, such as arrows, circular arrows, and boxes (see Fig. 8.3).

In this second application, the creation of the instructions can be completely done by domain experts. The expert can therefore select different 3D models from a table for every step and place, and rotate and scale them using hand gestures. For each step, a free text description can be added. Similar to the first application, the expert can document each step via pictures or videos with the support of the built-in RGB camera of the Microsoft HoloLens. The interface for viewing the instructions within the AAR application follows the same design as the CAR application. The simple 3D models are shown on the machine, as shown in Fig. 8.3.

Fig. 8.3 Abstract AR visualization of assembly step a) 1, b) 7, c) 11

8.5 Study

8.5.1 Setup and Apparatus

We collected the data based on the performance of an identical complex task. We instructed the participants to equip a Wafios RBV 35 bending machine with a set of bending tools using the pre-defined visualization types (paper, AAR, AAR+V, CAR, CAR+V). At the beginning of the task, no tools were mounted on the machine and the machine axes were set to an initial position, which did not change during the course of the study. The individual tool components were lined up on a table and individually labelled. The position of the tools was identical during each study.

The participants were given the task to mount the tool components on the machine according to the instructions. The instructions specified a certain sequence which was identical for all visualization types. The participants were instructed to follow the sequence, but they were allowed to go back and forth between the individual steps. In total, the entire process included eleven steps, each with one tool component. Tool component nos. 5, 8 and 11 include additional screws. No other tools (e.g. Allen wrenches) were required for the execution of the steps, as the screws only required hand tightening. This is slightly different in reality, but it makes the entire evaluation setup easier. Figure 8.4 shows the dependencies of the individual tool components. Some tool components do not have mechanical connections with others. For example, tool component no. 1 could also have been mounted in the last position. In contrast, it was necessary to install tool component no. 2 before no. 3. The situation was made more difficult by the fact that tool no. 4 could only be fitted correctly if nos. 2 and 3 were fitted correctly, see Fig. 8.4. This realistic conception of the task thus included complex steps that are difficult to perform without prior knowledge. We determined the complexity for each step according to various criteria (see for example (Radkowski et al., 2015)) on a 4-point scale from low to high complexity.

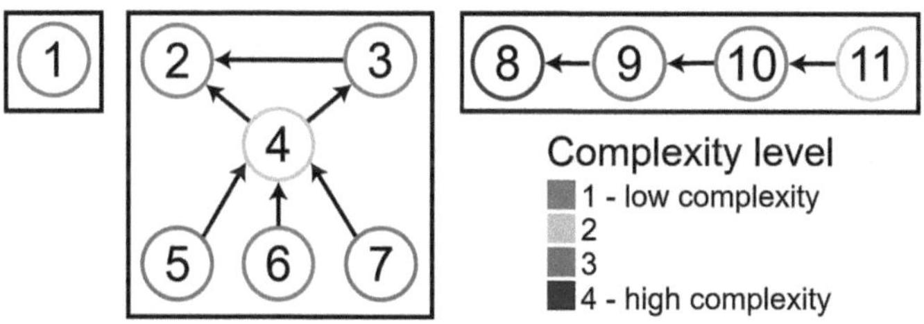

Fig. 8.4 Dependency graph of the assembly steps. Each box indicates an assembly group. There are no dependencies between the assembly groups and the assembly order could be changed. The arrows indicate the dependencies of the individual steps, e.g. step 5 depends on the correct assembly of step 4. The colour of the circles indicates the complexity level of the step

For each step, we created paper instructions and content for the AR applications. We selected and placed holograms for each step and recorded a short video showing the task execution from a first-person perspective. Within the AR applications, the name of the components and the holograms were shown. There were no additional textual instructions. Only for the AAR+V and CAR+V, further videos were provided. In the paper instructions we included a single picture of the mounted component and another picture of the component from the CAD data, as seen in Fig. 8.5. Additionally, the step number, name of the component and a one-sentence description were provided next to the picture. Each step was printed on one page in landscape orientation. All instructions were evaluated and approved by a mechanical engineer with domain knowledge.

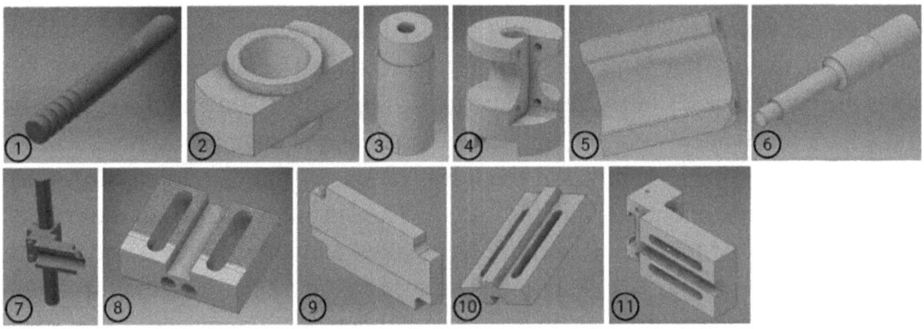

Fig. 8.5 Representations of the tool for each step

8.5.2 Data Collection and Processing

We collected several kinds of data during the execution of the study. The execution times of each individual assembly step were recorded by using an integrated logging function of the HoloLens applications. Specifically, the dwell time (including the time for watching the videos) of the participants in each step was logged and added to a total time at the end. The start and end of each step is indicated by pressing the next button in the main panel. For accurate measurements, e.g. to compensate orientation time in the beginning, we added one dummy step before and after the actual task. The entire procedure was also recorded with a video camera. The logged data were subsequently checked against the video and adjusted if the participants accidentally opened the bloom menu and did not know how to close it. In this case, we intervened and later removed this period from the total completion time. The errors were recorded manually and documented with pictures after the completion of the procedure. An error was counted when the position or orientation did not match the specifications of an instruction step. We did not distinguish the error type; however, subsequent errors were not counted if the relation to the dependent components was not broken. In total eleven errors are possible, one for each step.

We further collected additional data using four standardized questionnaires. We used the (1) Rating Scale Mental Effort (RSME) (Zijlstra & Van Doorn, 1985) and the (2) NASA-TLX (Hart & Staveland, 1988) questionnaires for measuring the task load, the (3) System Usability Scale (SUS) (Bangor et al., 2008) for measuring the perceived usability of the system, and the (4) User Experience Questionnaire (UEQ) (Laugwitz et al., 2008) for measuring the user experience.

To enrich our quantitative data, we additionally video-recorded the execution of the task and audio-recorded a follow-up discussion which was carried out as a closing interview. This discussion followed a predefined interview guideline. The audio recordings were fully transcribed and coded using a deductive coding system which was derived from the interview guideline.

8.5.3 Participants

We recruited 48 voluntary participants in total (7 female and 41 male) through social media and public mailing lists. Most of them (33) are undergraduate students of different disciplines (21 industrial engineering, 5 human-computer interaction, 6 mechanical engineering, 1 business administration). The remaining 15 participants are professionals in other industries (e.g., banker, journalist, soldier and software developer). We ensured that none of the participants had experience in operating the machine. Five persons had previous experience with setting up machines. We divided the participants into five groups by separating them equally according to their jobs and prior knowledge in mechanical tasks and setting up machines. From the participants in the AR conditions, 15 participants did not know about AR, 18 participants knew the concept of AR but had never used an AR application, and 7 participants had previously used some kind of AR (3 of these participants had previously used the Microsoft HoloLens).

8.5.4 Procedure

We began our study by informing each participant about the overall goal and the procedure. All participants signed a consent at the beginning of the test session informing them about their right to stop participating in the study at any time without any consequences. Furthermore, we informed the participants that the session would be video- and audio-recorded, but all recordings would be treated confidentially, and references would only be made in anonymized form. The participants had sufficient time to read the consent form and ask questions before signing the form.

Before we began with the actual task, we first calibrated the Microsoft HoloLens for each participant with the built-in calibration application. Next, we started one of our two applications and used a sample instruction for explaining the application and the interaction with the Microsoft HoloLens. The process lasted until we were sure that the users had mastered the most important gestures, especially the AirTap. Once this was achieved, we moved to the actual machine and we briefly explained the general task. We also showed the participants where the components were located and told them that they were completely on their own and would not receive any support. After the task was completed, we removed the Microsoft HoloLens, cleaned it for the next participant, and asked the participant to answer the RSME and NASA-TLX questionnaires. We then discussed the task with the participants near the machine and asked them to comment on the individual steps and rate them according to the perceived complexity. For this purpose, a 4-point scale was taken, where 1 corresponds to low perceived complexity and 4 to high perceived complexity. The participants were then asked to complete the UEQ and SUS questionnaires. Finally, the procedure was completed with a closing semi-structured interview. Overall, the procedure lasted an average of 70 minutes.

Of note, due to the ongoing COVID19-pandemic, we and the participants wore oronasal masks and maintained a minimum 1.5-meter (around 5 ft.) distance. In addition, the test area and gear were cleaned and disinfected vigorously before and after every session.

8.6 Results

With the data collected from the participants, we statistically analysed the values of the dependent variables. As the one-way ANOVA requires normal distribution and homogeneity of variance, we first tested for normal distribution using the Shapiro-Wilk-Test (Shapiro & Wilk, 1965) and for the homogeneity of variance using the Levene-Test (Tomarken & Serlin, 1986). The assumption of normal distribution was invalidated by the error rate, and homoscedasticity was violated by the RSME values. For the variables with normal distribution and homoscedasticity we used a one-way ANOVA for the analysis, the error rate was analysed with a Kruskal-Wallis test (Vargha & Delaney, 1998) and the RSME values with a Welch-ANOVA (Tomarken & Serlin, 1986). We further used the Bonferroni post hoc test (Tamhane, 1977) for the one-way ANOVA and Games-Howell post hoc test for the Welch-ANOVA (Table 8.2).

Table 8.2 Test results for normal distribution and homoscedasticity

Dependent Variable	Shapiro-Wilk-Test		Levene-Test	
	Sig. (p)	Result	Sig. (p)	Result
Task Completion Time	.102	OK	.499	OK
Error Rate	.008	NO	.293	OK
RSME	.461	OK	.001	NO
NASA-TLX	.631	OK	.140	OK

8.6.1 Task Completion Time

The first hypothesis (H1), which states that the mean values of task completion time differ significantly between the different visualization types, is not confirmed ($F(4, 43) = .194$, $p < .940$) by the one-way ANOVA. A closer look at the Bonferroni post-hoc test consequently shows that no combination of the dependent variables reveals a significant difference between the mean values of the execution times. The descriptive statistics show that the paper instruction required the least mean time for completing the whole set-up procedure with 646.5 s (SD = 237.8 s), followed by AAR with 675.0 s (SD = 171.6 s), and AAR+V with 677.9 s (SD = 149.6 s). The concrete visualization types both required more time, with 708.0 s for CAR (SD = 206.1 s) and 712.0 s for CAR+V (SD = 152.9 s) (Fig. 8.6).

There was a huge difference in the usage of the videos. According to the HoloLens log, the participants using AAR+V watched the videos more frequently than those using CAR+V. On average, the participants using AAR+V watched the videos 18.5 times (SD = 2.77) whereas participants using CAR+V only 11.3 times (SD = 6.59). Participant CAR+V-8 did not even watch a video once. CAR + V-2 stated: *"I watched the videos for details. Sometimes you did not even need the video because the task was so simple."* The time for watching the videos is included in the total completion time. However, there was no correlation between the number of videos watched and the task completion time.

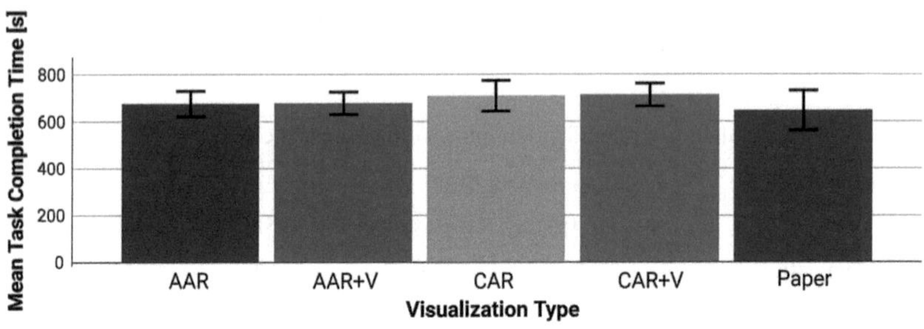

Fig. 8.6 The mean task completion time using the different visualization types. Error bars represent the standard error of the mean

8.6.2 Errors

The second hypothesis (H2) states that the mean number of errors differs significantly between the different visualization types. In this case the number of errors does not show a normal distribution; therefore, we performed the Kruskal-Wallis test. The results show that the mean number of errors differs significantly (Chi-square(2) = 21.671, p < .001). Descriptive statistics show that participants with a CAR+V visualization made only 1.7 errors on average (SD = 0.67). The average number of errors for paper and AAR+V was 2 (SD = 1.19 and SD = 0.66). Participants with AR visualizations without video had the highest number of errors. The average number of errors for CAR was 2.7 (SD = 0.82) and for AAR was 3.9 (SD = 1.1) (Fig. 8.7).

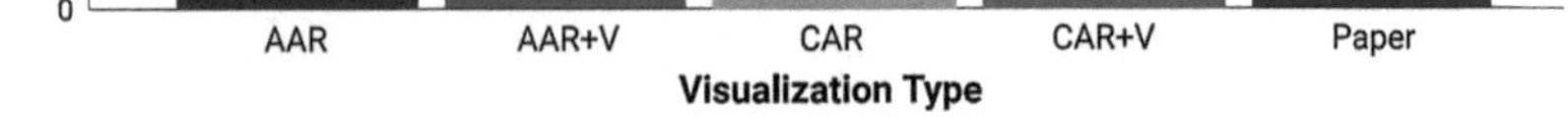

Fig. 8.7 The mean number of errors that were made during the task using the different visualization types. Error bars represent the standard error of the mean. The significance lines show the significance value of the pairwise comparison of the Kruskal-Wallis Test

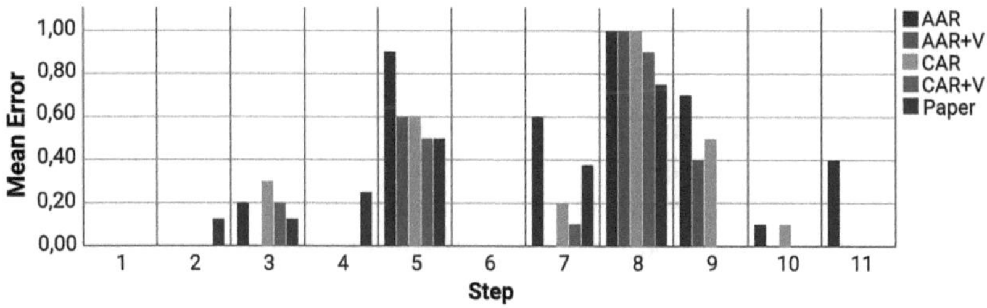

Fig. 8.8 Distribution of mean errors by step, grouped by visualization type

A closer look at the pairwise comparison shows that CAR+V and AAR differ significantly ($p < .001$) with a large effect size ($r = .912$). On average, participants that used CAR+V made 2.2 fewer errors than those using AAR. The same effect holds for AAR and AAR+V. Participants using AAR+V had an average of 1.9 fewer errors ($p < .005$) with a large effect size ($r = .774$). Even participants with paper instructions, with 1.9 fewer errors on average, had a lower number of errors and are therefore significantly ($p < .016$) different from the AAR instruction with a large effect size ($r = .742$). The results show that participants using abstract AR visualizations without video perform poorly in this setting, as they made a particularly high number of errors. The average number of errors associated with AAR+V is significantly lower compared to AAR.

The differentiated consideration of the error mean values related to the individual steps and the visualization types is shown in Fig. 8.8. We clearly saw that, on average, most errors occur in set-up steps 5 and 8. The number of errors reflect the pre-determined complexity of the individual steps. However, the perceived complexity was not in line with the pre-determined complexity, except step 1 and 6. Overall, the participants rated low complexity steps higher and high complexity steps lower. For example, step 8 was rated with an average of 2 by the participants versus our pre-determined complexity of 4. Furthermore, there is no tendency that indicates that videos lead to an overall lower perceived complexity.

8.6.3 Mental Effort

The third hypothesis (H3) states that the mean values of the indicators for cognitive load differ significantly between the different visualization types. For this purpose, we investigated the results of the RMSE and NASA-TLX questionnaires.

The RSME values did not show variance homogeneity and therefore the Welch-ANOVA was investigated. The results show that there is a significant difference between the groups (F(4, 21.325) = 3.080, p < .038). The Games-Howell post-hoc test does not show significant differences between the groups. Only between AAR and paper was the significance level almost reached with p < .055. Alternatively, the Bonferroni Post-hoc test was applied, which shows a significant difference between AAR and Paper (p > .010) with a medium effect size (f = .26) (Fig. 8.9).

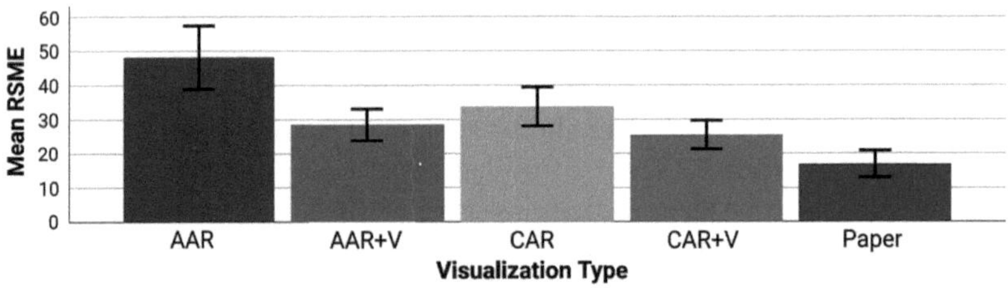

Fig. 8.9 The mean mental effort during the task using the different visualization types. Error bars represent the standard error of the mean

An examination of the descriptive statistics reveals that the RSME mean for using paper at 16.87 (SD = 10.99) is much lower compared to AAR at 48.15 (SD = 29.15). Both AAR and CAR result in a higher RSME value compared to paper instructions. When comparing CAR with 33.75 (SD = 17.91) and AAR, CAR tends to be slightly lower. The same applies to AAR+V with 28.55 (SD = 14.57) and CAR +Vwith 25.45 (SD = 13.33). These results show that videos lead to lower RSME values, which indicates less stress compared to set-ups without video.

8.6.4 Task Load

The NASA-TLX values show variance homogeneity and were tested for significance using one-way ANOVA. The test shows significant differences between the groups (F(4, 43) = 2,695, p < .043). In contrast, the Bonferroni post-hoc test shows no significant difference between the groups. Only CAR and CAR+V are approximately significantly different (p < .057) with a small effect size (f = .204).

The descriptive statistics provide a good indication that the mean value of CAR with 35.91 (SD = 13.59) and CAR+V with 19 (SD = 7.39) is different. This confirms the tendency of the RSME values described above that indicate that video reduces cognitive load. The mean value of CAR+V in the NASA-TLX is lower than the mean value of the paper instruction with 26.25 (SD = 10.95). AAR with a mean value of 34.33 (SD = 14.44) is similar to the mean value of CAR, but AAR+V with 28.91 (SD = 16.32) is higher than CAR+V (Fig. 8.10).

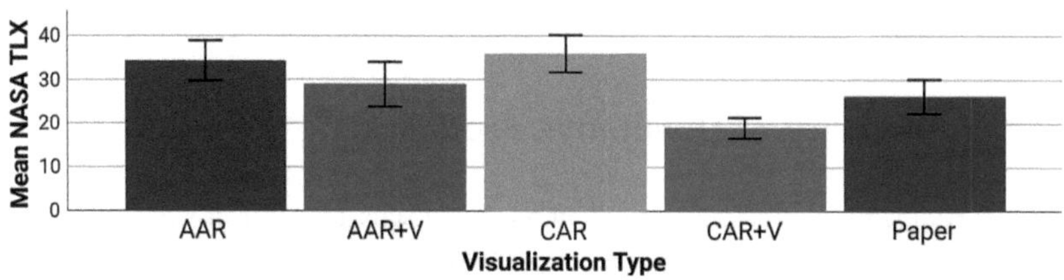

Fig. 8.10 The mean task load during the task using the different visualization types. Error bars represent the standard error of the mean

8.6.5 Usability and User Experience

The standard SUS-questionnaire according to Bangor et al. (2008) was used to evaluate the usability of the AR visualizations by the participants. The individual scores were combined to determine a mean value and were also tested with a one-way ANOVA. The preconditions of variance homogeneity and normal distribution of the mean values are given. The result of the one-way ANOVA shows no significance between the mean values of the individual visualizations ($F(3, 36) = .483$, $p < .696$). The descriptive statistics provide a mean value of 82 (SD = 12.19) for AAR+V. AAR has a mean value of 79.5 (SD = 9.26), which is only slightly lower. The same is the case for CAR+V with 78.25 (SD = 9.36) and CAR with 76 (SD = 13.5).

The UEQ was used to measure user satisfaction (Schrepp et al., 2014). After completion of the set-up task, the UEQ questionnaire was filled out by the study participants. For the analysis, the standardized evaluation procedure was used and the mean values were determined, as shown in Fig. 8.11. The results show no essential differences between the mean values.

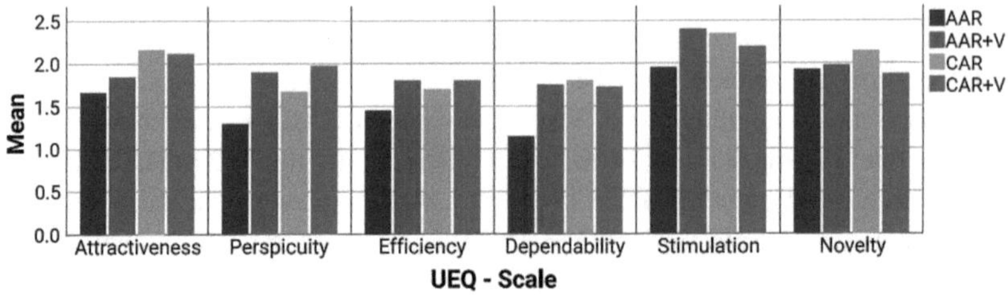

Fig. 8.11 The mean UEQ for the different scales grouped by different visualization types

8.6.6 Summary of Qualitative Feedback

Most participants expressed enjoyment in the discussion and closing interview even though half of the participants expressed discomfort related to the weight and comfort of the Microsoft HoloLens. The participants also stated they see high potential for the use of AR for providing assistance in the private and professional sectors. Participant AAR-4 was one of many that gave an example for private use: *"The application was very helpful to me and I can well imagine that it would also support me, for example, in assembling an IKEA cabinet."* Participant CAR-3 further explained: *"I found it much easier than the paper instructions you always get. Because there I don't understand anything anyway. When I see it virtually, I find it much easier."* Participant CAR-1 described a potential use case for the professional sector:

> *For example, for apprentices or trainees who do not have much experience, they learn how to do it properly. No one would have to stand next to them and explain it, instead they can do it them self, because they have everything in front of their eyes.*

There were also critical statements. Participant AAR-1, for example, said:

> *I noticed that because I was so concentrated on the device and the instructions, I didn't notice important details. You look less at the parts. I think when you work with the AR glasses, you turn off your brain a little bit.*

For similar reasons AAR+V-6 suggested *"I would have found it cool to get feedback from the system. For example, when you placed it correctly, that there is some audio feedback that indicates the part was aligned correctly."*

The participants using AAR had some issues with identifying the meanings of the abstract 3D models and therefore made the most errors. For this reason, Participant AAR-8 requested additional textual information, AAR-6 asked for concrete presentations of the components, and AAR-5 wanted to see videos about the execution of the tasks. Participants using the concrete AR application also had suggestions with regard to the information visualization. Participant CAR-3 suggested additional simple visualizations like arrows or circles for highlighting important details or positions. CAR-1 asked for the ability to hide the holograms after it became obvious where to mount the component because the holograms blocked his view on the real part. Multiple participants using the videos wished for more detailed videos, for example, with additional highlights of important details with marks or zooms.

Overall, the videos were perceived as extremely helpful by all participants. As AAR+V-1 put it: *"In the video I see how it is done and I just imitate it."* CAR+V-6 explained how she used the holograms and videos:

> *I used both equally, because they provided different information. So, the hologram for the position, there I did not look at the part itself and in the video, I was able to get more information about the exact look of the part and where the screws go.*

When we asked the participants, who had used the videos whether they could have completed the task without the video, they expressed skepticism and said that the videos gave them a feeling of safety and strengthened their confidence in their own actions.

8.7 Discussion

The goal of our study was to examine the influence the visualization of AR content has on complex working practices. We investigated if significant differences exist in (1) task completion time, (2) number of errors and (3) cognitive load between the visualization forms (Fig. 8.12).

Fig. 8.12 Comparison of the correct mounting of the component in step 5 and a typical error done by the participants. Left: AAR visualization with hint for attention to the chamfer. Middle: correct orientation. Right: the wrong orientation leads to a gap between the components

8.7.1 Type of Visualization has no Impact on Completion Time

The first hypothesis (H1) is not confirmed. The descriptive statistics show that AAR provides slightly better completion time values than CAR. These results are in line with results presented by Funk et al. (2015). A detailed analysis of our results (see Fig. 8.6) shows that the standard deviation of the task completion time for paper instructions is much higher than the standard deviation of the task completion time using the other visualization types. Participant CAR-3 suggested that paper instructions are often not very easy to interpret. This statement is reflected in the higher standard deviation. The results that CAR and AAR do not differ significantly in execution time support the findings of Blattgerste et al. (2018). However, the results of Radkowksi et al. (2015) differ from our results as Radkowski et al. showed significant differences for task completion times using different visualization types. Also, supplementing AAR and CAR visualizations with videos does not result in significant differences in the execution time. Comparisons of the results between the studies in general are limited, since the studies differ greatly in their design and implementation (Smith et al., 2020).

8.7.2 Concrete Visualization with Videos Promotes a Correct Execution

The second hypothesis (H2) is confirmed as there are significant differences between the individual visualizations with respect to the number of errors. Most participants using CAR performed better than the participants using AAR. Both CAR+V and AAR+V are better than CAR and AAR. In the case of AAR+V, using videos cut the errors in half. Based on these results, it is obvious that the effort to create CAR is worthwhile, as more detailed holograms lead to fewer user mistakes during assembly. AAR should only be used if holograms are combined with a video, however. Only in this case the abstract information of the hologram can be translated into useful instructions for the user (Funk et al., 2015; Yuan et al., 2008). A comparable result is presented by Smith et al. (2020), who state that in-situ videos in an AR environment are an effective way to display procedural instructions for assembly tasks. In general, it can be stated that videos complement the holograms and lead to a higher user safety during the execution of the tasks. The participants expressed appreciation of the holograms providing the position and orientation of the tool on the machine. They also stated that if they were unsure about the exact steps required, they watched the relevant video for further information and support.

A more precise differentiation of the results categorized by process step shows that most errors were made in set-up steps 5 and 8 (see Fig. 8.8). The components that had to be mounted in steps 5 and 8 are almost symmetrical and the correct orientation can only be recognized by a small edge on the tool (see Fig. 8.12). In these cases, both visualization types without videos perform poorly, but even though both had cues for the correct orientation. In the case of AAR, an arrow hologram was used to point out specific features (for example the chamfer on component of step 5). The CAR holograms represent the geometric features in the CAD models per se. As the results point out, the cues—especially in the AAR visualization—were not recognized by the participants. Some participants mentioned this aspect during the interviews by asking for a more detailed description of the tool position and orientation in step 5 and 8. We therefore suggest adding textual information for these specific details.

As Radkowski et al. (2015) have already shown, set-up steps of varying complexity have a particular effect on the number of errors. The complexity of a single set-up step can be determined, for instance, by the maximum number of orientations of the object to be mounted and the number of contact surfaces which have to be brought into line. Radkowski's et al. (2015) results show that CAR visualizations perform better by causing fewer errors. However, the hypothesis that CARs are better suited for complex set-up steps cannot be confirmed. Our results extend these findings by showing that CAR may cause fewer errors the addition of videos is essential when focusing on complex set-up steps. For the complex steps 5 and 8, the instructions with video perform better than the instructions without video. In addition, paper instructions are also quite suitable for complex steps (see Fig. 8.8). Participant AAR-1 speculates on possible reasons why pure AR instructions do not work so well. In his opinion, the holograms cause the user to be distracted and pay too much attention to the holograms, turning off common sense. Based on this, concrete suggestions were provided during the interviews. For example, participant AAR-8 requested additional written information for particularly critical set-up steps. In addition, the participants would have liked additional feedback on critical set-up steps (AAR+V-6). This suggests that the potential limits for both CAR and AAR were reached, since only the target state is visualized with no verification of the manual assembly process. This could be supported by additional sensory feedback components of the machine. An additional motivation for a subsequent sensory check is the partial lack of accuracy of the holograms. Especially for tasks in which exact placement of the assembly components is important, both CAR and AAR had limitations due to the hardware (Blattgerste et al., 2018). Our study does not provide any further insights regarding input interaction mechanisms for HMDs. The literature only refers in one place to the fact that different input interactions have no further impact on performance (Smith et al., 2020). However, our qualitative results highlight that participants emphasize further interaction with the AR-device in the form of direct feedback on the performed manual operation in the case for complex assembly operations or to use voice interaction to ease the general interaction with the applications.

8.7.3 The More Concrete the Visualization, the Lower the Cognitive Load

The third hypothesis (H3) is not confirmed. The Games-Howell post-hoc test could not identify significant differences between the groups; only the pairwise test of paper and AAR was close to significance ($p < .055$). The task load was the lowest with CAR. The Bonferroni post-hoc test revealed no significant difference between the group, only a trend that CAR+V performs better than CAR ($p < .057$). In their study, Blattgerste et al. also show significantly lower values for paper instructions (Blattgerste et al., 2017). Furthermore, AAR+V and CAR+V lead to a decrease in the cognitive load of the participants compared to AAR and CAR. We observed this tendency also in the task load; although they are not statistically significant, they are in line with the findings from the literature (Blattgerste et al., 2017; Smith et al., 2020) and warrant further research. The values for CAR and paper differ greatly between the mental effort and task load and they are not in line with our expectations. Participant CAR-1 reasons the lower task load by the autonomous learning environment that the AR instructions provide. The novices are therefore able to act on their own. In general, this is more successful with CAR compared to AAR and videos in particular reduce the task load considerably.

Both the results of the SUS and the UEQ indicate that there are no problems regarding usability and user experience. The SUS score is above the critical value 68 (Bangor et al., 2008) in all cases considered. The user experience is also rated above average for all visualization forms (Schrepp et al., 2014). The comparison between the visualization forms on the basis of descriptive statistics reveals that AAR is slightly lower than the others. The low values of AAR for "Perspicuity", "Efficiency" and "Dependability" can be traced to the fact that only abstract representations are not sufficient (see AAR-8, AAR-6 and AAR-5).

8.7.4 Trade-off Between Accuracy of CAR and Applicability of AAR

Currently, paper-based instructions are still status quo in many industrial settings (Engelke et al., 2013). However, AR-based approaches are increasingly gaining ground in this domain. As participant CAR-3 points out, AR instructions are generally much more accessible than paper instructions. This statement is not consistent with the results from the study and should therefore be judged as a subjective feeling. As our results indicate, the use of paper instructions is certainly justified in simple and complex settings. While in this study, we mainly focused on visualization, the creation of content must also be considered. Content creation requires several skills such as 3D modeling, computer graphics and animations, programming and expertise about the spatial registration of virtual elements and its tracking. Machine set-up experts often have little or no knowledge of creating AR content (Gattullo et al., 2019), which means that appropriate AR authoring tools are required to allow a quick and easy way to create content. Methods that involve generating AR content automatically are not suitable in view of the complex tasks (Fernández del Amo et al., 2018).

Instructions using AAR can be created directly with the Microsoft HoloLens and none of the listed skills are usually required, as predefined standard sets can be used. However, the results of the study show that the pure abstract representation is not sufficient for real world tasks, and too many errors are made as compared to CAR. Therefore, it must be weighted up whether concrete visualizations are needed, or whether abstract visualizations are sufficient.

8.7.5 Videos have a High Positive Effect on Abstract Visualizations

Although our study has shown that the addition of videos has a positive effect on both visualization types, AAR should always be combined with videos. The combination of AAR and video results in significant improvements and thus meets both requirements; easy and fast creation of AR content, and good performance when using instructions. As participant AAR+V-1 states, the video sequences are well suited to simply being imitated. Moreover, from the participants' (CAR+V-6) point of view, holograms and videos provide different information regarding the set-up task, so that both types of visualization complement each other and do not provide redundancy. Only the increased cognitive load when using AAR+V compared to CAR+V is a disadvantage. With regard to content, Gatullo et al. (2019) highlight the updateability of the instructions. Both AAR and video have good updateability if the AR hardware supports the creation of videos.

8.7.6 Take-aways for Designing AR-instructions

Our study shows that when focusing on minimizing the error rate, CAR visualizations are much more appropriate than AAR. When dealing with complex tasks, the error rate is often more crucial than the completion time. We therefore recommend using CAR to avoid errors. However, designers must consider the influence of the AR instructions on the users. Holograms can easily lead users to just focus on the holograms.

As adding videos as an additional resource, does not have a negative effect on the execution time, we recommend to integrate additional media like videos and photos or even text if holograms only cannot convey all important information reliable.

More strictly, we recommend for both visualization types that videos should always be considered as an essential part of AR-based instructions because videos lower the perceived complexity of a task, they lead to a reduced mental effort during the execution of a complex task and they can reduce errors.

We further recommend, whenever, 3D models of machine parts already exist and AR development expertise is available, consider using CAR or CAR+V. Creating CAR and CAR+V require additional effort in preparing the instruction, but for complex tasks this pays off. In the case that no 3D models are available or the creation of the models is too time-consuming, our results show that AAR+V has a performance comparable to CAR+V. The average number of errors is slightly higher, whereas the mean task completion time is lower.

8.8 Limitations and Future Work

Both the sample size and the set of participants pose a limitation in that the results are not fully representative. A larger sample size as well as a more representative set of participants with different domain knowledge probably would lead to further insights regarding the types of visualizations. To shed more light on the influence of video performance, it would be necessary to adopt a video-only condition in a similar study in the future. Further limitations are the slightly different interfaces of the CAR and AAR applications. The influence of the interfaces and a certain learning effect of the participants during the execution of the study were neglected in the evaluation. Although we assume that many of our findings can be applied to other complex tasks, we cannot say so at this time. Therefore, further research is necessary to investigate other fields of application and, if necessary, to make valid statements about the transferability of the visualization forms.

In addition to that we plan to focus on the effort of domain experts in creating the individual types of visualization and whether the effort is justified from their point of view. Although our study provided us with a scenario that was as close to reality as possible, our participants were all inexperienced in handling the machine. In future work, it will be interesting to investigate what effects the respective types of AR visualization have on experts in the field of machine set-up. For example, we assume that experts with significant previous knowledge need fewer concrete visualizations as they already know many aspects of the set-up. Other considerations for further work include investigating the effect of the HoloLens hardware and the users' experience with it. We will further consider multi-modal interactions like voice, hand gestures and gaze and dwell, depending on the task and environment. In addition, the level of detail in the holograms that can be placed on the physical machine to enable a precise representation should be evaluated (Blattgerste et al., 2018); this was particularly evident in the execution of step 8 in this study.

8.9 Conclusion

As the use of AR in industrial settings increases, it is important to understand how different visualization types influence user's performance. While previous studies typically either focus on comparing different media (Syberfeldt et al., 2015; Uva et al., 2018) as well as AR hardware types (Blattgerste et al., 2017; Funk et al., 2016) or using static apparatus' (Radkowski et al., 2015), we evaluate different types of visualization for the same hardware type for complex tasks in mobile settings. To accomplish this, we conducted a comparison study in which we identified the influence of the AR-based visualization types on the performance of complex machine set-up tasks. Both abstract and concrete visualizations have advantages and disadvantages. A designer should always make sure that the choice of elements fits the complexity of the task. Therefore, especially combinations of both visualizations types are useful and there is not only either or. We revealed different lessons learnt for designing AR-instructions using abstract as well as concrete visualizations as well as evaluation criteria.

To conclude our findings, we can state that CAR does not provide a significant advantage to AAR in terms of completion time. In contrast to this, CAR outperforms AAR regarding error rates, and videos lead in both cases to fewer errors -- especially for complex set-up steps. Overall, with regard to complex tasks, CAR is the better type of visualization. However, it generates additional work in the course of authoring which should not be ignored in such highly specialized application areas. The need to create time-consuming CAR content can be avoided by complementing abstract visualizations with videos, such that AAR+V is an efficient alternative to CAR or even CAR+V. AR glasses include various on-board features to easily create AAR instructions. If instructions will be created and updated by non AR developers on the shop floor, we recommend using these on-board features to create AAR or AAR+V.

Despite the limitations and the need for further research, our study provides a novel evaluation of different types of visualization on the same hardware and with the same media type for applicability in complex application settings. We not only contribute design spaces for HCI researchers, but we also provide concrete starting points for AR designers and hardware developers.

9 RetrofittAR: Supporting Hardware-Centered Expertise Sharing in Manufacturing Settings through Augmented Reality (JCSCW 2023)

Since almost the onset of computer-supported cooperative work (CSCW), the community has been concerned with how expertise sharing can be supported in different settings. Here, the complex handling of machines based on experience and knowledge is increasingly becoming a challenge. In our study, we investigated expertise sharing in a medium-sized manufacturing company in an effort to support the fostering of hardware-based expertise sharing by using augmented reality (AR) to 'retrofit' machines. We, therefore, conducted a preliminary empirical study to under-stand how expertise is shared in practice and what current support is available. Based on the findings, we derived design challenges and implications for the design of AR systems in manufacturing settings. The main challenges, we found, had to do with existing socio-technical infrastructure and the contextual nature of expertise. We implemented a HoloLens application called RetrofittAR that supports learning on the production machine during actual use. We evaluated the system during the company's actual production process. The results show which data types are necessary to support expertise sharing and how our design supports the retrofitting of old machines. We contribute to the current state of research in two ways. First, we present the knowledge-intensive practice of operating older production machines through novel AR interfaces. Second, we outline how retrofitting measures with new visualisation technologies can support knowledge-intensive production processes.

9.1 Introduction

Computer-supported cooperative work (CSCW) research has, from the beginning, focused on how knowledge and expertise are constituted in their respective practices and how they can be mediated through the use of technology. Indeed, CSCW

S. Hoffmann, *Knowledge Sharing on the Shopfloor*,
https://doi.org/10.1007/978-3-658-50133-4_9

has already contributed significantly to alternatives to the 'standard' knowledge management approach (see e.g. (Nonaka, 1994)) in and through a focus on knowledge as embodied in action. This 'practice paradigm' is by now prevalent in the CSCW community, and assumes the need for a detailed understanding of workplace practices prior to the development of technical artefacts to support expertise sharing (Schmidt, 2012). While early approaches around knowledge sharing focused mainly on artefact-centred repositories that encompass information production, storage, retrieval and reuse, more recent approaches focus on the ways that situated and socially contextualised knowledge is shared among knowledgeable actors and how it could be best supported through information technology, as 'expertise sharing' (Ackerman et al., 2013). These expertise-sharing approaches, often referred to as second-generation knowledge management, establish a perspective that embeds knowledge exchange as mutual learning in and through work practice. This relates to older discussions regarding the distinction between 'knowing how' and 'knowing that' (Ryle, 1945). One classic example is Answer Garden. This tool supports the development of an organisational memory by helping people to find answers to existing questions (Ackerman, 1998) or by locating experts within an organisation (Reichling, 2008). Recent research has demonstrated that new technologies have the potential to support expertise sharing in some radical ways (see Chap. 5) suggesting that AR and sensor technology, for instance, could result in a 'third generation of knowledge and expertise sharing research' (see Chap. 4). Such a move is quite distinct form orthodox views of knowlegde management insofar as they position themselves in relation to the 'embodied' rather than the 'tacit', and in doing so allow for a conceptualisation of knowledge as practice-based, contextually driven and evolving in realtime.

The discussions around expertise sharing become particularly relevant given the current upheaval in manufacturing industries. Many production facilities are presently being transformed by so-called cyber-physical systems, which intertwine physical and software production facilities that operate on different spatial and temporal scales. These systems interact with each other in ways that change contextually. Nevertheless, the supposition that technological innovation of this kind will necessarily transform all manufacturing environments is unwarranted. The manufacturing which takes place in many small- and medium-sized companies is usually based on *specialised* machines that have been in place for many decades. These machines were designed and built explicitly for the production of *specific* and *individual* products. Such machines cannot simply be replaced (both from an economic and organisational point of view) by modern cyber-physical systems.

The 'SME problem', so to speak, is compounded by the diminishing availability of special machine operators in these SMEs, who are gradually ageing and retiring and thus creating what was once termed an 'organisational knowledge' problem. 'Know how', put simply, is leaving the company. The experiential and embodied knowledges of long-serving employees often play an important role in production quality. In the context of Industry 4.0, then, we argue that the transfer the accumulated experiential knowledge to young professionals through improved contextualisation with the help of new technologies is a relevant and significant challenge for SMEs (Ludwig et al., 2016). In this paper, we report on a case study of a medium-sized company that is facing this precise situation. On the one hand, the company's machinery consists of old machines of proven reliability. On the other hand, the company must repeatedly replace its retiring employees—without losing the expertise acquired over decades. Specifically, we examined how legacy machines can be 'retrofitted' using modern lightweight augmented reality (AR) technology to prevent this loss of expertise and to support new employees in learning how to operate these specialised machines. With this paper we take up developments in the CSCW literature regarding expertise sharing by developing new mechanisms for storing machine settings as parameters. Visualization techniques associated with AR allow us to partially explicate knowledge in the form of parameters and to display it in situ in the working environment using AR technology. Chapter 5 shows that the processes described can be supported by AR, as the knowledge transfer model is essentially realised by the technical possibilities of AR. The goal in using this technology, then, is to initiate an expertise sharing process for complex production processes. Based on the findings of our detailed pre-study of the practices of workers in operating and configuring special machines, we designed, fully implemented and evaluated the AR-based HoloLens application, *RetrofittAR*. RetrofittAR enables relatively inexperienced machine operators to operate machines through an innovative interface. Our study clearly outlines how AR technology is beneficial for an expertise-sharing process on the shop floor in a real production environment, even when old machines without numerical controls are being used. The novelty of the paper is to present the results of a practical study with an AR retrofitting application to support expertise sharing and the recognition that the successful implementation of new technology in a manufacturing context, here an AR retrofitting application, may depend on an understanding of existing knowledge practices and their relationship to existing material infrastructure. Moreover, and unlike our work, little current research demonstrates the role of AR in supporting dynamic, real-time, operations.

9.2 Related Work

Our research combines three research discourses. The first focuses on human-centred manufacturing and the importance of expertise sharing. The second encompasses the appropriation of sociable technologies, and the third deals with novel interfaces for manufacturing settings in particular. Central search terms in the course of our literature search were: 'human centered CIM', 'organisational memory', 'augmented reality' AND 'expertise sharing' AND 'hardware', 'augmented reality' AND 'retrofit' AND 'industry', 'human in the loop' AND 'industry'.

9.2.1 Human-Centred Manufacturing

Various industrial revolutions over time have seen the development of new and different specialised markets. Because customers increasingly demanded individualised and customised products, the dynamics of production, were transformed. For this reason, new forms of manufacturing, work design and innovation became necessary to cope with highly dynamic markets. In this constantly evolving field, computer integrated manufacturing (CIM) has been challenged by alternative approaches around human-centred manufacturing systems. These human-centred forms of organising manufacturing processes, which are substantially based on human skills, experience and competences, are often called 'anthropocentric production systems' or 'human-centred CIM' (Badham, 1991; Brödner, 2007). There is an extensive literature which discusses the characteristics of human-centred manufacturing and the challenges associated with it (Corbett, 1990; Peruzzini & Pellicciari, 2017). As Grandi et al. (2018) point out, this requires a focus on *"performances, aesthetics, reliability, usability, accessibility and visibility issues, costs, and many other aspects."* (p. 702) and hence, a transdisciplinary approach is necessary (Grandi et al., 2020). A more limited literature has focused on knowledge sharing in the manufacturing context (Li et al., 2019) who reference the urgent need for *"a holistic framework for identifying and accommodating individuals' needs and expectations of relevant data, information and knowledge"* (p. 380).

The radical change of management strategy from Tayloristic models to a human-centred, expertise-based approach that promotes continuous competence is ongoing and, represents the beginning of a new phase of industrial development. In contrast to previous manufacturing revolutions, which generated significant productivity growth, the new industrial revolution arguably generates new innovation dynamics dependent on the sharing of knowledge and expertise. The exchange of expertise presents itself as a complex social interaction process (Brödner, 2007), whereby developing an organisational memory (OM) is likely to be an essential driver of a company's success. The concept of OM reveals how organisations use and maintain knowledge in various forms (Bannon & Kuutti, 1996). However, OM is neither simply stored nor unproblematically accessed (Ackerman, 1998; Ackerman & Halverson, 1999; Randall et al., 1996). Knowledge and expertise are not subject to rigid structures but rather constitute living and changing content in practice, which in turn carries special implications for possible support technologies (Bannon & Kuutti, 1996).

9.2.2 Hardware-Based Expertise Sharing

With regard to supporting expertise sharing within manufacturing settings, a number of standard procedures exist, including for instance the use of plans and shift protocols (Carstensen et al., 1999) or coordinating meetings (Schmidt, 1991). However, when focusing on the practice of manufacturing itself, these mechanisms usually have fallen short because a large number of contextual circumstances are neglected in respect of actual machine operation (Lewkowicz & Liron, 2019). These contextual factors, such as the detailed nature of the materials, the actual condition of the machine (Bowers et al., 1995) or even the current room temperature, can play a decisive role in manufacturing.

Ludwig et al. (2017) focused primarily on these contextual and environmental factors to design a hardware-based approach for collaboratively dealing with this complexity in practice. Using Turkles' (2002) original concept of sociable technologies, they show how Internet of Things technologies can offer operators new possibilities for communicating, documenting and sharing hardware usage practices. Sociable technologies offer three dimensions for understanding a machine's functionality and sharing its practical uses: (1) the *internal* dimension, which provides the user with information on the inner workings of a machine, its current status and the functioning of the complex component structure; (2) the *spatial-material* dimension, which provides detailed information about the machine's location and environment (e.g. room temperature and brightness), so that possible relationships with the internal context can be derived; and (3) the *task-* and *process-related* context, which provides information on the position in a production chain or purpose and objective of machine use (Ludwig et al., 2014).

When interacting with specialised machinery, the operator is the central unit (Cimini et al., 2020). With the advent of human-centred manufacturing (see Sect. 9.2.1), new technologies are usually introduced in manufacturing settings with the aim of supporting users rather than replacing them. Although physical demands are changing due to the use of new technologies, cognitive abilities are also potentially increased by assistance systems (Romero et al., 2016). This interaction between operator, technology and organisation (Dregger et al., 2016) leads to an augmented knowledge base, in which human abilities and technical possibilities complement each other (Engelbart, 1988). Specifically, the ability to interpret data and derive decisions through automated collected data paves the way for extensive expertise sharing (Romero et al., 2016).

9.2.3 Supporting Hardware-Based Expertise through Augmented Reality

Historically, hardware has often only poorly provided information about machine behavior, especially in the case of machines from the first to third industrial revolutions. Written manuals, checklists (paper-based and digital), status updates and so on, provide information but only in a static manner. While modern machines with various sensors provide for much more sophisticated monitoring procedures, traditional interfaces are significantly more restricted in their ability to provide real-time visualizations (Ludwig et al., 2019). In the context of human-centric manufacturing, it is apparent that effective operations need exactly this kind of reactive, dynamic, visualization and AR is considered to have the potential to provide interfaces to display real-time feedback during the operation of machinery. Such interfaces can also be considered as second-level functionalities which not only allow the user to interact with the machine but also explain and support the interaction thus, inter alia speeding up accurate decision-making (Jasche & Ludwig, 2020; Ludwig et al., 2017; Stevens et al., 2009). Sensing the machines' behaviour and interconnectivity are prerequisites for these interfaces (Al-Maeeni et al., 2020).

Guerreiro et al. (2018) have demonstrated how retrofitting old machines (adding components not included with the original machine) can meet the requirements of modern manufacturing. The technology used in the retrofitting process is aimed not simply at automating the machines but also ensuring that operators receive better feedback from the process. Here, appropriate visualisation of process data is an essential interface between the machine and the human being. The individual design of the retrofit application must also consider the specific user requirements (Ramakers et al., 2016). In the case of retrofitting measures, smart glasses are increasingly used as a visualisation tool, which permits rapid data interpretation and immediate decision-making (Al-Maeeni et al., 2020; Guerreiro et al., 2018). In addition, the spatial recognition capabilities of an AR device can be used to recognise machines, suggest suitable components for a retrofit and visualise them directly. In addition, subsequent actions such as the purchase or sale of such components as well as instructions for assembly or disassembly can be supported with the AR device (Mourtzis et al., 2020).

Research in the area of smart glasses has shown that workers in industrial settings can readily be supported in assembly tasks (Baird & Barfield, 1999; Bhattacharya & Winer, 2019; Caudell & Mizell, 1992; Dey et al., 2018). Research on systems which support workers in performing assembly tasks by displaying the necessary information at the relevant times, goes back almost 30 years (Caudell & Mizell, 1992). In comparison to other instructional media, such as written paper or videos, AR and its multimodal form of feedback provide speed advantages in the assembly task's execution (Baird & Barfield, 1999; Fiorentino et al., 2014; Henderson & Feiner, 2011; Jetter et al., 2018). In addition, AR-based instructions allow for a higher level of detail and thus further reduce cognitive load (Bresciani & Eppler, 2009; Paas et al., 2003; Tang et al., 2003). When combined with cyber-physical systems, smart glasses may support the exchange of knowledge and experience between experienced and inexperienced production workers (see Chaps. 4 and 5). However, most AR studies in industrial contexts focus on more static situations like assembly tasks or maintenance and there is little research on using AR as support during the actual operation of a machine.

Augmented Reality clearly requires authoring processes, scripts or instructions in order to reduce training and other overheads (Ramirez et al., 2013; Roberto et al., 2016; Seichter et al., 2008). These authored instructions, in the main, aim to support reoccurring and static situations like assembly or maintenance tasks. Further work focuses on automatic algorithmic authoring techniques which are based on existing data such as product information or prerecorded videos of a workflow (Fernández del Amo et al., 2018; Petersen et al., 2013). Techniques which are based in particular on video recorded workflows are useful because the data can not only be used to automatically create instructions, but can also be used to analyse the users' actions and to provide feedback on accuracy. In unforeseen and dynamic situations without prerecorded instructions smart glasses remain useful. Most smart glasses can transmit the field of view of the support seeking person to a remote expert over the internet. The expert can now see the situation and can provide support via speech or by placing virtual objects like arrows in the local environment (Ludwig et al., 2021).

9.3 Research Approach and Application Field

Our review of the literature suggests that there are no current examples of AR technology being used to enhance knowledge sharing while operating and visualising setting parameters on (older) special machines. There is a clear need, we argue, in settings where the operation of machinery requires specialised knowledge, and where the passing on of knowledge and expertise cannot be guaranteed, for dynamically available information to be provided in support of manufacturing practice in situ. Visualisation techniques hold evident potential. We addressed this specific, but important, research gap in CSCW by examining how AR technology might support operators' expertise sharing in adjusting and handling machines continuously during manufacturing. We introduced, for the first time, the possibility of using AR to retrofitted legacy machines by designing new interfaces for expertise sharing. The novelty of the paper therefore, and as indicated, consists of the results of the evaluation of an AR retrofitting application in a real production environment. By doing so, we aim to answer the research question: *How might an AR application be designed to retrofit legacy machines to foster expertise sharing for knowledge-intensive production processes?*

To answer our research question, we examined manufacturing processes in a small German company. Although the increased use of numerical controls for production machines, in particular, can result in high manufacturing precision at fast working speeds (Kumar et al., 2018; Schlegel et al., 2018), specialised machinery that was constructed for specific products decades ago is not equipped with such technological accessories. The machine park of the company we investigated consists of several specialised machines, which are operated by experienced and long-standing employees, and whose expertise has accumulated over decades. However, this type of manufacturing process comes with a risk. If an experienced employee retires, the company must hire a new employee who can handle the machine. Because such employees are difficult to find, the company struggles with prolonged idle time, which can quickly problematise its ability to survive in today's market.

The company we investigated produces tank heads for apparatus and plant construction and specialises in the metal spinning production process. The process is a single-stage, tool-bound forming process that offers high productivity due to short cycle times. The company owns several specialised spinning machines. Some of the machines are equipped with computerized numerical controls, and others are without such controls. The machines produce heads with different geometries and various materials. The basic machine structure is shown in Fig. 9.1.

Fig. 9.1 Metal spinning machine

Figure 9.2 shows the machine's production process set-up: (1) a rotating tool provides the final contour of the head, (2) the head is pressed against the tool by a press roll (3) with a certain pressure, (4) which is applied to the press roll along the x-axis and thus produces plastic deformation. In addition, the pressure roller moves at a certain speed v along the z-axis to the edge of the head. The x- and z-axis and their parameters p and v are individually controlled by the machine operator for each head by a rotary control.

The combination of the tool shape, the raw material in the form of a circular blank and the parameter settings for the x- and z-axis result in the final product of a head. Thus, the metal spinning process comprises several variables that must be assessed and analysed without the support of a numerical control.

For more insight into the company's structures and practices, we conducted a design case study, which typically consists of three phases (Wulf et al., 2011). The first phase entailed an empirical analysis of the existing practices and usage tools. The second phase comprised analysing results for an iterative and participatory generation of design ideas and solutions. In the third phase, the appropriation of the technological artefact in a real setting, its potential redesign and the artefact's potential future impact were documented.

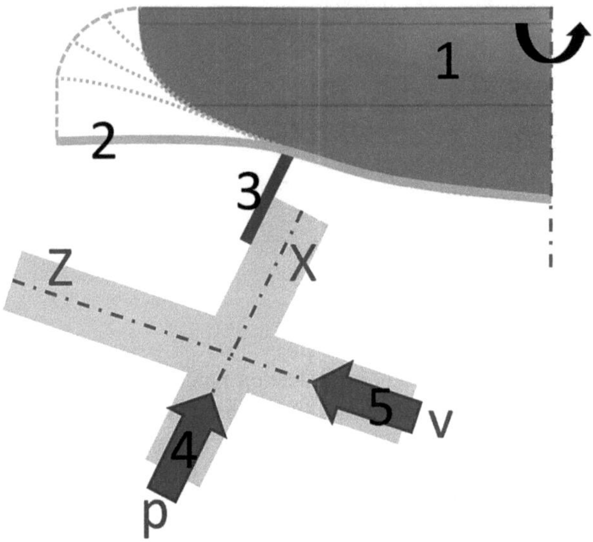

Fig. 9.2 Machine set-up, (1) tool, (2) head, (3) press roll, (4) x-axis with pressure p, (5) z-axis with speed v

9.4 Empirical Study

9.4.1 Methodology

To examine current practice, we organised and documented eight workshops, observed machine set-up and spinning processes through eye tracking and conducted four semi-structured interviews. The workshops were held together with various representatives in different organisational positions (see Table 9.1; W1–8). A total of seven representatives from production management, work preparation, machine set-up and technical management participated in the workshops. The workshops were documented through notes made by the researchers and jointly prepared working papers that were discussed and improved in the ensuing workshop. In concrete terms, the contents discussed in the workshops were recorded in note form. After the workshop, the contents were formulated and recorded with sketches (examples are Fig. 9.2 and Fig. 9.4). These contents were verified at the beginning of the subsequent workshop with the core team or distributed to the workshop participants and then improved in a joint session. In workshop 7, a scheme for data collection was jointly elaborated and converted into an Excel document. Subsequently, the documents produced were progressively filled with data by the company (see Fig. 9.6 sheet of paper). Overall, a summary document—a kind of manual—was written, which was improved in several iterations and finally approved by the participants and the management as a complete project documentation. This collection resulted in a 250-page document with all details of the spinning process during the one-year cooperation recorded.

Besides a general briefing, we wanted to observe and quantify the machine operator's activities and perspective during the process (E1–2). With the help of eye tracking, fixations (points of view that are looked at closely) and saccades (rapid eye movements between fixations) were recorded and analysed. We equipped an experienced operator with the Tobii Pro Eye Tracking Glasses 2. To ensure accurate tracking, the machine operator went through a short calibration phase. Afterwards, the operator started the spinning process. Because eye tracking can be operated wirelessly, the glasses did not restrict the machine operators' work, providing us with real-time and detailed insights. Thus, we could observe how the employees interacted with the machine during the process without disturbing it. Practical limitations meant that only two recordings were made with one individual who had the capacity to carry out the complex processes and simultaneously record and explain his rationale with the glasses. Nevertheless, eye tracking allowed us to collect data that the machine operator might not otherwise have been able to directly communicate. To evaluate the collected data, we created visuals in a form typical for eye-tracking data: (1) the gaze plot was used to show the sequence and duration of fixations during the process and (2) the heat map also made it possible to identify which areas of the machine and the machine interface were viewed particularly frequently and at what time. The operator was also asked to comment on their actions.

Table 9.1 Overview of workshops during the pre-study and design phases

Workshop [W] Interview [I] Eye-Tracking [E] *Duration*	Scope	Participants (gender / work experience)	Role	Guiding questions and *materials*
W1 *4 h*	Project definition	P1 (m / >30 yrs)	Technical Manager	What challenges does the company have to face? *Field notes of the conversation.*
		P2 (m / >20 yrs)	Commercial Manager	
W2 *7 h*	Process analysis in general	P3 (m / > 15 yrs)	Production Manager	What does a production process from incoming goods to delivery look like and what are the challenges in the processes? *Field notes of the conversation and process models.*
W3 *7 h*	Analysis of the operations spinning process	P4 (f / >10 yrs)	Work Planner	What are the exact operations of the spinning process? *Detailed work plans and sketches.*
		P5 (m / > 30 yrs)	Foreman	
W4 *3 h*	Analysis qualification profile	P4	Work Planner	Which qualifications are needed for which operations? *Development of a detailed qualification matrix.*
		P5	Foreman	
		P6 (m / < 1 yrs)	Machine Operator	
W5 *7 h*	Technical foundations	P1, P4, P5, P6		How are the machines technically equipped? *Field notes of the conversation.*
W6 *5 h*	Disturbance variables	P4, P5, P6		What influence do disturbance variables have on the process? *Wear measurements of the press roll and calculation of the influence on the parameters.*

(continued)

Table 9.1 (continued)

Workshop [W] Interview [I] Eye-Tracking [E] *Duration*	Scope	Participants (gender / work experience)	Role	Guiding questions and *materials*
W7 *10 h*	Detailed settings	P4, P5, P6		What do the pressure and speed curves look like for the heads? *Excel sheets with pressure and speed parameter.*
W8 *12 h*	Design and technical realisation	P4, P5, P6		What information must appear in the operator's field of view and how should the connection between HoloLens and machine be technically implemented? *Detailed design concept and technical documentation of the WebSocket connection.*
		P7 (m/ > 10 yrs)	Maintenance Staff	
E1 *2 h*	Operating procedure	P5		When and why are which settings made and what does the machine operator observe? *Transcripts and eye-tracking recordings.*
E2 *2 h*	Operating procedure	P5		
I1 *50 min.*	Spinning process and design	P3	Production Manager	Guide questions about (1) personal details and the current focus of activity, (2) the production process under consideration and the existing challenges and (3) the possible ways in which visual guidance could be provided in the execution of the production process. *Transcripts.*
I2 *50 min.*	Spinning process and design	P1	Technical Manager	
I3 *40 min.*	Spinning process and design	P8 (m / > 10 yrs)	Master Craftsman	
I4 *40 min.*	Spinning process and design	P9 (m / > 15 yrs)	Logistics Specialist	

After the workshops and eye-tracking recording, we conducted semi-structured interviews. The interviews were recorded, transcribed and analysed and lasted on average for 45 minutes. The interviews were based on an interview guide with 14 questions for the Master Craftsman and Logistics Specialist. The five central questions were as follows:

- What makes the forming process difficult to learn?
- How did you learn to operate the machine? What training levels did you go through?
- What makes a successful forming process? How does a good result look like?
- What do you use as a guide when you operate the machine, for example increasing or decreasing the pressure? What role does your experience play in this?
- Are there differences between the various heads in terms of the forming process and the degree of difficulty?

The interview guide for the Production Manager and Technical Manager comprises 12 questions. The five central questions were as follows:

- What do you consider to be relevant data for operating the machine?
- What specifications/content would you like to make/teach?
- What are the disruptive factors (Why is process-safe production a challenge)?
- Please describe the connection between school-based and in-company training and the acquisition of process knowledge!
- Everything is allowed and possible: What would you do/provide to make process-safe production possible?

Based on the empirical data, we derived some design challenges (see Sect. 9.5), which led to our concept for supporting hardware-based expertise sharing through AR (see Sect. 9.6). We implemented our concept as the HoloLens prototype RetrofittAR, the performance of which was later evaluated.

9.4.2 Results of the Empirical Study

As previously mentioned, the company is in competition with several other companies and has to maintain productivity levels while managing costs. Thus, the company tries to differentiate itself through a special product portfolio. Additionally, it offers low prices, which presuppose a high degree of economic efficiency on the production side and the use of highly specialised machinery. Both aspects require that there is a huge amount of expert knowledge deployed in the specific production practices. We examine these aspects in detail below.

(i) *Practical Expertise Sharing—More Important than Theoretical Knowledge Transfer*

The operators play a decisive role in successful production, as they must be able to operate the specialised machines reliably (W1, P1). The challenge is that this reliability is based on implicit knowledge and experience gained over a period of time. Therefore, the production process's success relies almost entirely on individual operators. There is a large risk, as one interviewee suggested: *'At the moment, everyone who retires with us takes all the knowledge they have acquired in 20–25 years on the machines with them. Because it's not recorded or written down anywhere' (I1).*

Relying solely on school-based training is problematic because content adapted to specific company production processes is not available. The school-based part of the training, in particular, did not fit the company's requirements because its specific machinery was not included in school-based provision (W1, P2).

> *When we send the young people to basic training, filing is part of the curriculum. This does not add any value to our forming process. Rather, the trainees must learn what physical properties and forces are involved in the forming process. (I1)*

The company, therefore, tried to tackle this challenge through a long-term, dual training strategy. In addition to the theoretical content, which is taught during school-based training, the company fostered the transfer of practical content. However, it turned out that the significant level of error generated during the training process was unacceptable. Put simply, no mistakes are allowed.

> *The forming process is simply based on learning-by-doing. Gaining experience, gaining experience. And this experience is gained over time. And when you're faced with productivity and time pressure, those are two factors that are difficult. (I4)*

Under these conditions, it is of particular importance that the machine and its components is adequately explained, as this is the only way to ensure actual learning-by-doing within the actual production environment. I3 outlined:

> *First of all, it is important to know the machine and its characteristics. What you produce with the help of the machine. So that the new employee understands this. Then the control panel itself is important. What do the buttons mean? Where do I have to adjust what? But this static content can easily be transferred. (I3)*

I1 highlighted the importance of the expertise gained onsite during the operation of a machine:

> *Every article has a certain diversity. They may only be nuances, but you have to recognise the nuances in the production process on the manually controlled machines and react immediately. A cause–effect relationship must be established. Now this and that happens on the part, and I have to turn the wheel. Another part may have different adjustments. (I1)*

As a result, a knowledge gap was identified which had to do with inadequate theoretical training content and the highly contextualised manufacturing processes. The company, therefore, established a training phase for new employees in which they would be accompanied by an experienced foreman. The experienced employees withdrew from their day-to-day business, thus creating the capacity for in-house training activities. However, this in-practice teaching had some shortcomings:

> *From my perspective, the learning environment seems to be a big problem. We're training people on the equipment in a very loud and dirty environment. You have practically no learning workshop or closed learning area. You are in the production hall, in the dirt, right next to the noisy plant. And we don't manufacture any training parts either; these are all parts that are sold. (I1)*

To better plan and track staff training, the qualification levels of in-house training within W4 were jointly recorded and documented, as no corresponding documents were currently available: *'We also don't have any documents or tools that could be used to do this' (I2)*. Thus, the training's success was highly dependent on the people involved. Furthermore, in addition to the technical expertise, the ability to train new personnel was of decisive importance:

> *It is not up to every employee to train the employees on the machine. One can do it; the other can't. One can do it well; the other can do it badly. One person may misunderstand or take it the wrong way. It must be someone who passes it on sensibly. Not just anyone can do it. (I3)*

(ii) *Determining Deviations During the Manufacturing Process is Highly Contextualised*

During the production of tank heads, deviations between the planned products and the actual products sometimes occur. To determine the cause of these deviations, it was necessary for the machine operator to compare the head's current state with the target specifications. The head's target geometric specification is determined by drawings and standards. These standards included the head's written dimensions which instruct the operator which characteristics must be measured while doing the quality check (W5, P4). These pre-specified and written characteristics were labelled in documentation as hard quality characteristics. In addition, the surfaces and overall visual impressions were constantly checked by the operator using so-called soft quality characteristics (W5, P5). The operator either detected deviations after production by checking the head's circumference or by determining the head's behaviour on the machine at an early stage during the production process through observation (W5, P6) (see also Fig. 9.3): *'I determine the quality by closely observing the head on the machine. I pay attention to the gap between tool and head and to the fluttering movements of the head edge' (E2, P5).*

Fig. 9.3 Operator observes the spinning process in detail; left: head flutters; middle: the gap between tool and head; right: crimping of the head

The results of our eye-tracking study clearly showed that the operator observes the behaviour of the head on the machine in a highly focused manner at each stage: *'I must always keep my eyes on the head. This is the best way to detect deviations' (E1, P5)*. These observations allowed the operator to intervene during the production process if needed. Thus, they constantly compared the actual situation with a target specification and derived the control deviation. When a head was completed, the circumference was also measured. If the measurement was outside the tolerance, the head might need to be reworked or scrapped (W3, P5).

(iii) *The Real Expertise can be Seen in the Adjustment of the Parameters*

The operator typically uses the deviations described in the previous chapter to define the settings or manipulate input parameters. Our workshops revealed that observing the production process required considerable in situ expertise. Moreover, the parameter adjustments required extensive expertise acquired over many years and, above all, a constant check of adjusted parameters and desired effects in practice. We conducted a precise elaboration of the detailed settings requiring manipulation during the seventh workshop. First, the head was divided into different areas. However, these areas depended to a significant degree on the size and geometry of the head. We, therefore, developed a uniform definition of the areas together with the operators. Figure 9.3 shows that at least three phases occurred during a head's production. In the early production phase, the operator focused on the fluttering of the head. In the further course of the process, the operator focused their attention on the gap between the head and the tool, as well as the head's outer edge.

> *From the beginning, the first step, that is pure experience. That means that when I have a 3 mm sheet in front of me, I start with a pressure of 30, for example. That is pure experience. As soon as I start pressing, I have to watch the head very closely. That means I have to see how the sheet is behaving and increase the pressure according to the observed behaviour of the sheet. (I4)*

I2 expanded on this statement by pointing out several important parameters: *'From my point of view, there are three parameters to consider: pressure, speed and rotational speed. The pressure and speed vary along the radius of the head, and the rotational speed is always constant.'*

The interviewees told us that how the pressure is adjusted is of particular importance: *'The pressure must be increased slightly. Caution is required' (E1, P5)* or as P9 explained: *'The pressure must be carefully increased in several stages'*. We, therefore, developed the head plan shown in Fig. 9.4.

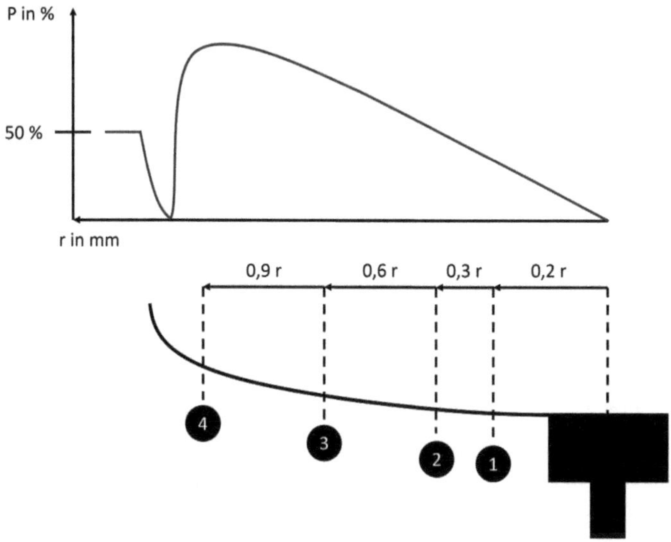

Fig. 9.4 Variation of the pressure profile along the radial direction of the head

To enable the machine operators to adjust the machine, using the machine's three main control variables: (a) the pressure of the x-axis, (b) the speed of the z-axis and (c) the rotational speed of the tool. For example, the following instructions resulted from observation and comparison with the setpoint. *'If the head flutters too much, I'll have to relieve the pressure' (E1, P5).* As the process progressed, it became more important to increase the pressure so that *'there is no gap between the head and the tool' (E1, P5).* Both settings referred to the pressure of the x-axis (blue line in Fig. 9.4). As the degree of deformation increased along the radius of the head, the pressure of the x-axis needed to be increased. This increase needed to be entered manually by the operator, via a rotary control on the control panel beside the machine (W7, P5).

As we previously mentioned, the exact course of the blue curve in Fig. 9.4 was different for each article to be produced. Therefore, the settings were recorded relative to the following characteristics (W7):

- Geometry and surface of the head
- Material of the semi-finished product
- Production machine

Any combination of features created new and very specific pressure profiles of the head and required that the machine operator know the settings for all feature combinations. In this case, the operator needed to have about 70 pressure profiles at hand. The former machine operator I4 described the relationships:

> *The bulge from the head is practically the pressure on the roll. The surface of the head is affected by the speed. The parameters are directly related to the machine and the raw material. Also, the speed and pressure are directly related. (I4)*

This complexity, consisting of the product-specific dependence between radial position and pressure was further increased by disturbance variables, such as wear and tear of the tools and the characteristics of the raw material. The results of the empirical study indicated that these pressure adjustments had to be changed by up to 30% due to disturbance variables (W7, P5). As I1 explained, *'On the one hand, the raw material ... we cannot guarantee that one component is really like another. On the other hand, the machine, where wear and tear cause changes in the condition of the tools' (I1).*

(iv) *Summary of the Empirical Results*

Our empirical study showed that the machine operator plays a crucial role in current manufacturing processes when handling specialised machines. Their expertise and training have a significant influence on the success or failure of the production process. The absence of an automated control system made this particularly relevant. Technically speaking, the machine operator took the place of this controller by performing sensory tasks, determining control deviations and defining manipulable variables. Inexperienced machine operators usually do not have such expertise. However, current strategies like the intensive training of inexperienced machine operators, as we have shown, fall short due to inappropriate theoretical training content and the highly contextualised manufacturing processes.

9.5 Design Challenges

Based on the empirical findings, we derived challenges and design implications that guided the later implementation of the HoloLens application *RetrofittAR*. The existing ways of operating the machines posed several challenges in relation to technical support for the practices of machine operators. On the one hand, the existing machinery was indispensable for economical production but, on the other hand, the machines did not convey relevant real-time data to the operator during the production process. When coupled with insufficient machine operator training, it became apparent that inexperienced operators required the machine's operations to be made visible during its actual operation. As the company's production requires specialised machines, the instructions had to be adapted individually to each machine. Due to the lack of qualified machine operators, the instructions needed to be easy to understand and easy to follow during the production process. The instructions also had to contain the essential single-setting parameters and be designed to support machine operators' practices. Furthermore, the instructions needed, as far as possible, to be integrated into the existing machine infrastructure, so that a retrofitting of the existing machinery would be possible.

As Sects. 9.4.2(ii) and 9.4.2(iii) show, the skills of machine operators played an essential role in production. They observed the quality of the output during the production process and directly initiated appropriate measures by adapting the setting parameters. Thus, placing all instructions directly in the operator's field of view was deemed important. Here, the required input parameter (see Fig. 9.2) must be visualised and coupled with machine data. Specifically, a fusion of input parameters with the machine data (e.g., speed or pressure) allow a deeper understanding of how to operate the machine appropriately. By visualising these values, the machine operator no longer needs to constantly look at the machine control panel to see the actual setting of the two parameters mentioned and is thus able to maintain the necessary head quality observation.

Because the parameters set vary along the head's radius, it is necessary to display them according to the head's geometry. For this purpose, a connection between parameter visualisation and the actual position of the machine axes is needed. Therefore, the target geometry of the head must be placed directly on the machine itself, including the parameter specifications. This visualisation enables a continuous comparison of the real machine data with the visualised specifications that can be carried out by the operator. However, for older machines, this was not a simple matter, as the positions of the machine axes and further machine information were not provided digitally.

Because the empirical study showed that several heads were produced on a single machine, we determined a visualisation of each head and the parameter specifications that must be adapted in real time. This is accompanied by the adjustment of the given parameters using the disturbance variables. As our results in Sect. 9.4.2(iii) show, the influence of the disturbance variables depended on the degree of wear and the characteristics of the raw material.

Table 9.2 summarises the design implications based on the empirical findings and links to the implementation presented in Sect. 9.6. We derived all design implications directly from the findings of the workshops, interviews and eye-tracking sessions presented in Sect. 9.4.2.

Table 9.2 Summary of empirical findings, design implications and implementation

No.	Empirical Finding	Design Implication	Technological Implementation
Socio-Technical Infrastructure			
1	Existing machinery consists of special machines and requires precise knowledge of how to operate the machines in practice (see 9.4.2(i)—I4).	The operation of the machine must be instructed and individually adapted to the existing machines.	Implementation of software adapted to the machine to display the machine settings (see Fig. 9.5 'Default parameters for the spinning process').
2	Essential basics are neither on a theoretical nor on a practical level part of the professional training (see 9.4.2(i)—I1).	Simple and realistic presentation of the instructions is needed for quick learnability of the basics and correlations between machine and operator.	Use of AR technology for the presentation of learning content directly at the workplace when operating the machine (see Fig. 9.9 for an example how the operator can focus on machine).

(continued)

Table 9.2 (continued)

No.	Empirical Finding	Design Implication	Technological Implementation
3	Personnel-intensive training phases lead to high training costs and few lasting qualification successes (see 9.4.2(i)—I4 and—I1).	Person-independent presentation of the training contents and support of the operators in the competence development are based on the recorded training material.	Implementation of a software based on a mobile AR technology for the representation of the important knowledge contents consisting of location-related parameters and the head geometries. Enable asynchronous knowledge sharing by capturing parameter settings from experienced operators using an Excel template and then converting them into AR instructions in the app (see Fig. 9.6).
4	The machine infrastructure is outdated, and yet economical production can be achieved with appropriately qualified operating personnel (see 9.4.2(i)—W1, P1 and 9.4.2(i)—I1).	Existing machines and adapting technology are used for representing instructions to the existing machines (retrofit).	Use AR to enable retrofitting of the existing machine with purely virtual content and instructions (see Fig. 9.5 for a general overview of the general AR-interface).
Highly Contextualised Expertise			
5	The machine operator ensures the quality of the production by intensive optical observation of the production process on the machine (see Fig. 9.3).	Compressed interface for all relevant information directly on the machine, so the operator can continue the necessary practice of observation.	Virtual geometry of the head is projected directly onto the real head. Setting parameters (current parameters and predefined parameters for speed and pressure) refer directly to the real machine (see Fig. 9.8).

(continued)

Table 9.2 (continued)

No.	Empirical Finding	Design Implication	Technological Implementation
6	Settings of the machine parameters by the operator vary along the radius of the head (see Fig. 9.4).	Visualisations of the settings must refer to the geometry of the head. The head must be virtually mapped.	The geometry of the head is calculated with the help of the virtual head generator (see Fig. 9.6) and projected directly onto the machine with Vuforia (see Fig. 9.8). Parameters are placed on the appropriate segments with ToolTips depending on the geometry (see Fig. 9.8).
7	A set of features describes a head and, to a certain extent, defines the settings for the machine parameters (see 9.4.2(iii)).	Depending on the characteristics, the instructions must be visualised specifically for each head / article.	The data model is built specifically for each head and consists of geometric data of the head and the setting data for the machine (see Fig. 9.6 and Table 9.3).
8	Disturbance variables lead to an influence on the setting of parameters (see 9.4.2(iii)—W7, P5 and 9.4.2(iii)—I1)	Recording of the disturbance variables enables correcting of the setting instructions according to the disturbance variables.	Adjustment of the disturbance variables according to the detected wear and automatic correction of the default parameters according to the disturbance parameters (see Fig. 9.7 'Main menu').

9.6 Concept

To tackle the aforementioned design challenges, we designed an AR concept that enabled the machine operator to observe the head's behaviour during the production process and, if needed, to make appropriate adjustments. Our empirical work, along with the eye-tracking, showed the importance of the continued monitoring of the machine by operators. Thus, we focused on using a head-mounted display to visualise the relevant information directly in the operator's field of view. At the same time, this allowed the operator's hands to be free to perform the necessary process steps.

We created a visual representation of our concept, shown in Fig. 9.5. The segments of the radius were mapped to the actual physical head so that the operator could see the current radius range, including the parameters for pressure and speed. These virtual objects were always aligned with the position and angle of the operator's view. We also displayed the default values for pressure and speed. Coupled with the virtual segmentation of the head, we highlighted the place where the operator must manually put in the values when the physical press roll of the machine touched the virtual segmentation. In addition, there was a live visualisation of the actual machine parameters. This visualisation always displayed the current speed and pressure. Our concept not only gave the machine a new way of representing the configurations but also made the monitoring of head behaviour easier.

Fig. 9.5 General concept of the AR-interface: (1) Virtual segmentation of the head along the radius, (2) Default parameters for the spinning process, (3) Live data of the machine

9.7 Implementation

We implemented the concept as a Microsoft HoloLens application called RetrofittAR. The aim was to provide the machine operator with the relevant information about the machine, the article and setting options. RetrofittAR was implemented with Unity. We used the Mixed Reality Toolkit (MRTK) because it already offers basic assets for lower-level tasks, e.g. for user interfaces. We mainly used the MRTK version 2 and extended it with some legacy assets of the MRTK version 1/Holotoolkit assets.

As the data within our concept was closely related to the nature of the physical and spatial environment, we used the image and object tracking Software Development Kit (SDK) Vuforia 8.3.8. Vuforia allowed us to scan image targets with the HoloLens camera to precisely position our AR images. The image targets were positioned directly on a flat and somewhat protected place on the machine body. Due to the poor lighting conditions in the production hall, tracking was very occasionally unreliable and deviations of up to 5 cm could occur. Therefore, a handle was implemented that allows manual correction of the positioning of the AR images. The user is now able to compare the virtual representation of the head with the real head and correct accordingly with the handle.

The requirements of our application meant we had to integrate different relevant data types and sources with RetrofittAR. Figure 9.6 shows the different data types and structures of our application. To implement the data about articles, the team first had to create databases with parameters for speed and pressure. We therefore created corresponding Excel sheets for each article. Within each Excel sheet, the operator listed the pressure levels, pressure profiles, speed and rpm for each article. Our findings had indicated that personnel-intensive training phases were largely unproductive due to the production environment and the low level of prior knowledge (see Table 9.2). Hence, an asynchronous knowledge transfer was initiated. For this purpose, the experienced machine operator had to fill Excel sheets with his preferred parameter settings and these were then implemented as scriptable objects in the Unity development environment (see Fig. 9.6) and visualised as default parameters in the ToolTips (see Fig. 9.8). The geometries of the heads were usually standardised. We, therefore, stored the basic dimensions and the exact shape of the head geometry for each article according to the construction guidelines. Thus, we were able to create each head geometry virtually on the HoloLens. The geometries could be quickly determined using the stored formula, even for new head dimensions.

For the machine's live data visualisation, we connected the machine's rotary control, signal lights and pressure indicator to a Raspberry Pi, which in turn provided the data via a WebSocket connection. The number of verticies of the virtual head and the amount of data from machine to HoloLens via the WebSocket connection were designed so that the performance did not fall below 40 fps.

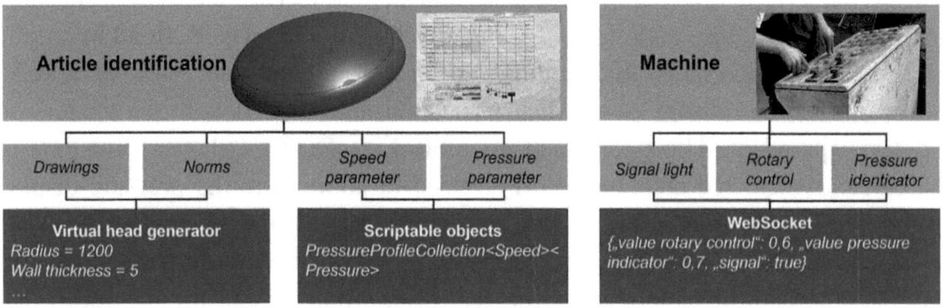

Fig. 9.6 Data structure of RetrofittAR

The following data structure emerged (see Table 9.3), which was necessary for the productive use of our application.

Table 9.3 Data structure

No.	Class	Description
1	<Article>	The base of the data architecture. It defines a combination of <Material>, <TorisphericalHead>, at least one <Machine> and at least one <PressureProfileCollection> associated with <Machine>.
2	<Material>	The Material of the head. It contains data about quality (hardness).
3	<Machine>	The Machine which is used to produce an <Article>. It contains information about Abrasion, Speed and Pressure.
4	<Torispherical-Head>	The geometric shape of the Article. Always a torispherical head, but its dimensions vary: Based on the editable fields ArchRadius, CornerRadius and WallThickness, more measurements are provided by this data container.
5	<PressureProfileCollection>	A List of at least one set of instructions (PressureProfile) to be executed, dependent on the progress of the spinning process.

The freely manoeuvrable AR image with the main menu allows the user to select articles from a multi-page list with the known article numbers (see Fig. 9.7). After selecting an article, its geometric dimensions are displayed. If this article can also be produced on another machine and parameters for the settings have also been stored, the user can then select an alternative machine and the corresponding parameters for this machine are displayed. In addition, the wear of the roll and batch fluctuations of the material can be set with two sliders. According to the settings of the disturbance variables, the default parameters on the ToolTips adapt. The freely manoeuvrable Live Data UI showed live data of the currently selected machine (see Fig. 9.7). In addition to the pressure display of the ToolTips, the current pressure value was also displayed on the Live Data UI by means of a circular scale which was designed according to the scale on the machine control panel. The same scale was implemented for the speed display. Furthermore, a single machine signal was transmitted and displayed, which indicated when a certain position on a machine axis had been reached.

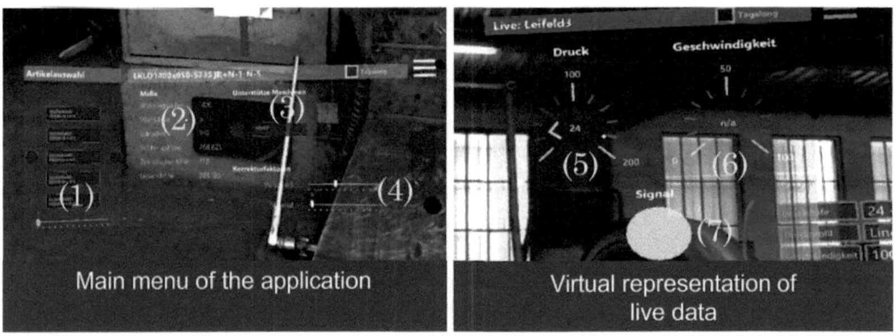

Fig. 9.7 Main Menu: (1) Article selection, (2) Dimension of the head according to the selected article and calculated with the virtual head generator, (3) Machine selection, (4) Adjusting the disturbance parameters; Virtual representation of live data: (5) Live data speed, (6) Live data pressure, (7) Live signal

The Torispherical Head UI was placed as an overlay over one of the respective real machines and tools in three dimensions with Vuforia and the MultiVuMark package. Based on the drawings and standards supplied, the head was calculated and finally rendered on the HoloLens. A small section of the head was displayed and rendered from one side only so that the operator had a clear view of the head (see Fig. 9.8). In addition to the head, indicators were placed on relevant spots of the currently displayed 'TorisphericalHead' shape, segmenting the head along the radius. As shown in Fig. 9.8, the relevant spots were at 0,2r; 0,3r; 0,6r and 0,9r.

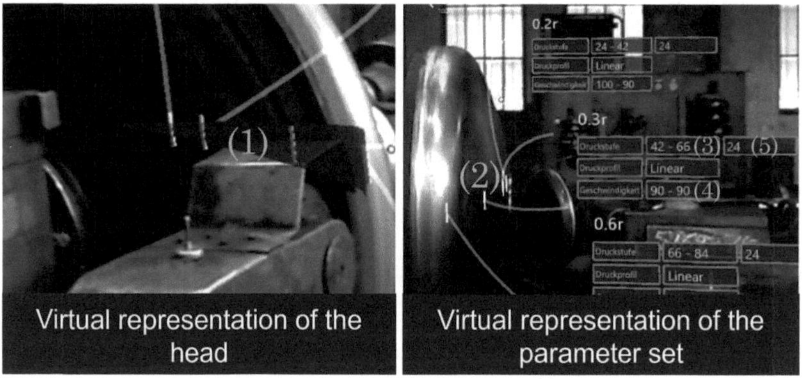

Fig. 9.8 Torispherical Head UI: (1) Rendered part of the virtual head, (2) Indicators for segmenting the head along the radius, (3) ToolTip containing predefined parameter for pressure, (4) ToolTip containing predefined parameter for speed, (5) ToolTip containing the current parameters for pressure

The indicators were connected to the ToolTips by a curved line. The ToolTips contained the PressureProfileCollection and showed the predefined parameters for pressure and speed at the relevant spot. Next to the predefined parameters, the current pressure value transmitted by the machine over a WebSocket connection was displayed to the right (see Fig. 9.8). These virtual objects were billboarded so that the position of the object was always aligned with the position and angle of the machine operator's view.

9.8 Evaluation

9.8.1 Methodology

To evaluate RetrofittAR, we conducted an evaluation study with 10 machine operators aged between 25 and 60 (average age of 35) (see Table 9.4). None of the participants had previous experience with AR technology. Our evaluation consisted of a task-based in situ evaluation of our prototype in a real production environment. We followed with an interview with each participant.

Table 9.4 Evaluation participants

#	Procedure	Participant	Role	Professional experience
T1	unaccompanied	P5	Foreman	> 10 years
T2	unaccompanied	P10	Operator	approx. 5 years
T3	accompanied	P11	Operator	no experience
T4	accompanied	P12	Operator	no experience
T5	unaccompanied	P13	Operator	approx. 4 years
T6	accompanied	P3	Production Manager	no experience
T7	unaccompanied	P14	Operator	approx. 6 years
T8	unaccompanied	P15	Operator	approx. 2 years
T9	accompanied	P16	Operator	approx. 2 months
T10	accompanied	P17	Operator	no experience

During the in-situ evaluation, we first explained the HoloLens and its interaction modalities. Afterwards, all participants independently went through the HoloLens's built-in calibration app. Calibration was necessary so that the HoloLens fit individually into the participant's viewport. We then instructed the participants to launch our app and select the corresponding production article from a list. Although we had previously stored the data of 60 production articles within the application, the evaluation was performed only using four of them. We asked the participants to place all freely placeable AR images according to their preferences. Subsequently, the production process started, and the participants were advised to produce up to ten similar heads with the help of the application.

Our evaluation was split into two modes. In the first mode, an accompanied evaluation was conducted. In this case, an inexperienced machine operator wore the HoloLens and articulated the setting values that were visualised by

RetrofittAR to an experienced machine operator. The experienced machine operator set the values exactly as specified and only intervened if a collision was imminent. This procedure was implemented to ensure work safety, as the inexperienced machine operators were operating the machine for the first time. In the second mode, a completely independent evaluation was carried out. The machine operator performed the evaluation without external assistance (see Fig. 9.9).

Fig. 9.9 Accompanied evaluation (left) and unaccompanied evaluation (right) of the HoloLens application

During our evaluation, we were particularly interested in the use of RetrofittAR in the real production environment and the difficulties that arose. The philosophy behind the evaluation was derived from Twidale et al.'s (1994) 'situated evaluation'. Therefore, qualitative methods were applied which allowed us to draw conclusions from the use of the technology in the real production environment. Experts and novices were asked to describe their views on the usefulness of the application. The participants were instructed to 'think aloud' (Nielsen, 1993). Due to the noise level in production, statements could not be recorded, and only notes of the comments made by the operators were taken. Immediately after the in situ evaluation, we conducted a semi-structured interview with the participants. Our questions focused on the following topics:

- Information content of the visualisation.
- Interaction between real machine and virtual content.
- Placement of the information.

- Interaction possibilities and usability of the application.
- Integration of the glasses into everyday work.
- The extent to which the AR images represent knowledge-relevant data for the experienced employees.
- How the integration of the machine values into the virtual interface was perceived.

9.8.2 Results of the Evaluation

(i) *Display and Placement of AR Images' Decisive Influence on the Timely Processing of the Illustrated Parameters*

The participants emphasised that the AR images had to be moved and placed freely when needed (T2, T3). This functionality allowed them to keep the relevant values in view, even if the operator moved. At the same time, the participants were also able to observe the physical machine, which is one of the main advantages of AR:

> *You can observe everything on the machine and in the environment. These are not VR [Virtual reality] glasses, where you are simply isolated from the real environment. So, you still have everything in view and can place your windows as you need them and always have everything in view. (T2)*

The placement of the Live Data UI was especially important, as these values had to be continuously observed:

> *I put the pressure display directly over the tool and then I could always see how the pressure was. You have all the lines on the tool of the machine and next to them boxes with the values that show how much pressure you have to set. (T2)*

However, because the AR images of the live values could be moved freely, the operator had the choice of where to display the values, which challenged inexperienced operators. Thus, they expressed a preference for the pressure values of the machine, at a minimum, to be placed directly next to the reference values: *'The pressure values of the machine should actually be right next to the target values. So, I can compare both values directly and do not have to look at another place' (T6).*

The virtual line's location marking on the tool (see Fig. 9.8) was found to be accurate and appropriate, and the machine operator could see exactly where they needed to adjust the pressure and speed values according to the target values:

'So the lines definitely match the real tool. I followed the lines and adjusted the parameters accordingly' (T3).

Nevertheless, during the evaluation, it became apparent that with the small head diameters, the lines for the virtual representation of the head (virtual lines) were very close together at the beginning of the spinning process. The operator had to react quickly by increasing the pressure values at a precise time. As a result, the operator found it difficult to map the virtual target values with the real position of the machine in a timely fashion, losing the overview.

> *I wanted to set exactly what the glasses were supposed to do, but then I couldn't do it anymore. The lines were too close together because the diameter of the head is very small. Then I continued producing the head without this information. (T8)*

Thus, simply displaying additional and more detailed information was problematic, as the operators found it difficult to process too much information. *'So, the more information, the better. However, the more the information, the more difficult it is to process in practice' (T6).*

To simplify the interface, one operator suggested visualising only selective information: *'For me, there were too many markings constantly visible. If the marks would gradually appear and then disappear again depending on the position of the pressure roller, only the acutely important marks would be visible' (T7).*

(ii) *Using AR as an Upgrade or Supplement Highly Depends on the Expertise*

The experienced machine operators had difficulty growing accustomed to the new interface, as they found some information to be redundant and were distracted from the actual observation of the head's behaviour on the machine:

> *I concentrated on the glasses and also perceived the information. But for me, it was a strange feeling. I don't know how to explain it: I actually know what I have to do because I have been doing it for a long time. It was difficult for me to concentrate on the glasses and ignore how I had done it before. (T1)*

T6 described the experience as follows: *'It is decisive how I increase the pressure. If I increase the pressure, I must immediately pay attention to the reaction and behaviour of the head. Therefore, I will not rely completely on the glasses'.* T6 also indicated that the AR glasses are only a support for the operator and their value lies mainly in ensuring that the human is provided with the necessary information for setting the control parameters. For this reason, some operators thought

RetrofittAR should only be regarded as a supplement, depending on experience. T8 explained:

> *I think it's perfect for training people. You don't have to constantly look at the machine panel. All the values are collected on the interface, and at the same time, you have the head in view. Especially for the beginning, it is really good. For the people who have been working on it for a longer time, for them, it is not upgrading but supplementing. For people who are doing it for the first time or for people who are just getting started, it is definitely an upgrade. Because you save yourself the trouble of drawing lines on the desk or making other markings you can really have a view of the tool or the head and learn what you need to pay attention to. Instead of looking away from the machine every time to look at the panel and check what pressure is set – you have it right in front of you. You have all the information you need right in front of you. And that's definitely an upgrade, especially for training. (T8)*

Based on the existing data, T8 also pointed out that the interface should be seen as a supplement because not all external influences can be represented in the target values and therefore the visual control of the head's behaviour still had to be the main focus:

> *If the batch of the metal melt is changed, the entire material behaviour during the spinning process changes as well. Then the pressure is sometimes too high or sometimes too low. Both are problematic. So, the eye has to control itself. (T6)*

However, when focusing on inexperienced operators, T9 provided an important insight by stressing how sometimes one could gain a deeper understanding of the process:

> *Well, I've been working here for several years now, and from time to time I asked my colleague: What are you doing there now? What are you actually looking at here? The interface on the head made me realise which details he pays attention to and how the process works. (T9)*

RetrofittAR allowed the machine operator to perceive an almost undiminished environment and thus maintain communication with colleagues: *'And you can also talk to your colleague when something is going on somewhere' (T4)*. There are situations where glasses are not necessary, and the machine operator does not need the data provided by the interface. *'When changing the material, the glasses can be flipped up so that you can see more' (T5)*.

Fifty years span the machine and the introduction of smart glass. The operators initially had problems thinking away from pure hardware and towards a data-oriented view. Once this perspective was adopted, operators recognised the potential of AR as an additional interface for the machine that supports the production process:

Well, my opinion is that glasses and machines don't really fit together. The machine is from the '70s, the glasses from 2020, but from a content point of view, it makes sense to wear the glasses when operating the machine. (T1)

(iii) *Enabling Expertise Sharing Directly at the Physical Machine*

To support on-the-job learning and build up knowledge about the machine, it was important that the operators had a constant view of the machine's current values and 'behaviour':

You have the head and the machine in focus all the time, and you don't have to look at the machine console again and again. And that way you can start learning. In other words: To combine the virtual of the glasses with the real haptic of the machine. So that in the end, you get to be able to produce the head – I would say – by feeling. (T8)

Although RetrofittAR could make learning on the machine easier from the operator's point of view, there was a risk that inexperienced operators would rely too heavily on the application, thus preventing the actual building of expertise. T9 suggested:

The only risk I see is that people might try to learn by heart and not react to the behaviour of the head. That the operators don't take the glasses as a support, but that they now say: I take the glasses now and I know what it is written, and I don't have to do anything else. Then a new batch of material comes in, and then the values are again not 100% correct. You just can't rely on it 100%, but you rather have to take it as a guideline. (T9)

The lack of experience becomes evident when deviations or problems occur. Gaining this complex process knowledge is part of the efforts of inexperienced machine operators: *'You want to know what you are doing. You don't just want to press buttons; you also want to know when, how and why' (T3).* T3 described the exact content that he received through the AR visualisation:

For me, it is definitely a help because I am inexperienced. And I think it makes sense to get support as someone who is inexperienced in using such a machine. And if you have such a help, especially in the beginning, when you first have to get to know the

> *machine and learn when you have to increase pressure and when you have to decrease the pressure again, until when you have to keep a certain pressure and when you fold the head. I find that helpful, absolutely. (T3)*

RetrofittAR can even act as a mediator between experienced and inexperienced users. This role emerged during classical training situations in the company (consisting of an expert who explained and implemented the training and a novice who listened and watched). Questions that often arose could be better answered by using AR:

> *When I watch the experienced one during production, I always think: Why is he turning there now? Why is he spinning fast now? But this way, I understand the process more precisely. If you didn't have this insight before, with the help of the glasses, you can now understand it more easily. (T10)*

Even beyond the initial training phases, for some it made sense to use the glasses, especially when they were not sure how to handle a certain dimension of the head because they had never produced it before.

> *These glasses are not designed to be worn for 8 hours. You use them for the first 5, 6, 7 or 8 heads. After that you know: Here to this area,* I have to turn there, until then I have to turn there, until then I have to turn there. In some cases, the following situation can occur: At 2 o*'clock on the night shift there's no one there, and you have never produced 1250 × 5 head dimension. (T2)*

Thus, the glasses can create an autonomous learning environment from the operator's point of view, which partially allows the novice to learn the basic settings autonomously: *'The operator is taught the basics in a relatively compact time: pressure, speed, feed: 'And after that, he can teach himself many things, not everything, but a large part of it using these glasses' (T8).*

(iv) *Usability Issues*

Wearing the glasses was not always a comfortable experience, though this varied considerably from operator to operator. One user found that*'the glasses fit well on the head and can be adjusted well' (T10),* while others felt*'a feeling of pressure on the forehead and temples' (T5).* The interaction options were perceived as basically precise, although*'it takes several minutes to get used to them' (T2).* The menu navigation of the prototype was perceived as occasionally cumbersome due to the division of the contents into individual windows:

> *I had to operate via two different windows, which initially confused me. It would have been better to integrate all content in one window. In the end, however, only a few settings had to be made, which is good in this industrial environment and everyone can manage the operation. (T1)*

Due to the limited number of settings (article selection, wear setting and connection setup to the WebSocket), the menu structure was considered easy to use and T1 estimated that all colleagues could cope with the operation of the application.

Due to the difficult lighting conditions and the low illumination of the working area on the machine, the machine operators found the AR images too opaque:

> *It is not very bright in the production hall. It was really hard to see how the material was touching the tool. (T8)*

T8's statement confirms that observing the machine is of paramount importance (see Fig. 9.3). However, by adjusting the brightness of the HoloLens displays and reducing the size of the virtual head, problems could be remedied quickly and observation of the machine could take place without difficulty.

From a usability point of view, the integration of the machine displays into the AR interface was perceived as very positive. T7 states: *'The interface is complete and all information is visible on the machine'*. In general, no inconsistencies were found by the operators.

9.9 Discussion

The results of our evaluation demonstrate how the practices and skills of operators inform their attitudes towards RetrofittAR. More experienced workers had habitual ways of doing things and, to a certain degree, found the application to be unnecessary for their work practices. In particular, with the real-time processing of information and the execution of measures based on the visualised information, an information overload turns out, for them, to be problematic (Hiltz & Turoff, 1985). As our evaluation shows, experienced operators find the visualised information redundant (Kalyuga et al., 2003), which may lead to an information overload during the process's execution. The experienced operator simply preferred to rely on their own assessment and visual inspection of the process rather than trusting the given values.

Conversely, the inexperienced machine operator does not have this experience and is, therefore, grateful for the target values, even though these values can only provide reference points. Thus, the AR display is particularly suitable for inexperienced operators because the interface both provides a visual of the target values and allows the observation of the process. RetrofittAR provides inexperienced machine operators with the ability to perform the process independently and thus they can gain important practical experience and expertise in situ. This ability is important, especially against the background of the lack of theoretical training because all subsequent training content can be based on this expertise. Particularly, the semi-automation of expertise sharing in an actual production environment can positively affect expertise sharing processes (see Chap. 5). As we point out above, in a highly competitive environment, it is important to avoid interrupting normal production, and further to avoid unnecessary errors in production. Our evaluation demonstrates that the application helps in fulfilling these objectives.

The results of the empirical study showed that users pay considerable attention to the fact that AR content adapts to the actual situation. We, therefore, implemented a dynamic head geometry creation based on standardised design specifications. This implementation means that any dimensions of the heads could be generated through RetrofittAR in real time without having to create models externally in a design software environment. As Bannon and Kutti (1996) indicate, the visualised contents must adapt to the actual tasks in practice. The results of the empirical study show how very process-specific contents and representations are necessary when providing instructions for use with machines of the kind we were dealing with (see Table 9.2). Existing software solutions on the market are oriented towards the display of standard geometries such as boxes or arrows and do not offer the possibility of displaying process-specific setting data (Vuforia Expert Capture or Microsoft Remote Assist) or only support remote scenarios (Microsoft Remote Assist). Existing solutions simply do not meet the requirements for context specific information as described in Sect. 9.4.2(iv). On the other hand, the geometries and data must be made available to the software. Many of the authoring tools require extensive expert knowledge for the creation of the AR content (Nebeling & Speicher, 2018). To mitigate this issue, the virtual head generator (see Fig. 9.6) was implemented, which automatically generates the geometries when a new article is created. The input parameters were packaged in scriptable objects (see Fig. 9.6), which can also be easily created. One limitation is that the data in the current software have to be carried out in the Unity development environment, and although no programming knowledge is necessary for this, Unity must be operated. At this point, however, the process can be further simplified in the future on the basis of the virtual head generator and the scriptable objects by implementing an end user oriented authoring environment.

We argue that AR technologies are especially suitable to foster expertise sharing processes in a fast-changing knowledge environment where real-time production routines need to be maintained. They are suitable, not just because data and knowledge content can be adapted quickly and comprehensibly, but because the user must react to these changes and can build up new expertise more quickly and efficiently. In our opinion, the reason for the support of this AR technology transformation is the fusion of the AR images with the real environment. Additionally, the possibility—as shown in the evaluation—arises that default values and the process's live reaction to these values can be assessed by providing effective feedback through the glasses (Burke et al., 2006; Shah et al., 2012; Tang et al., 2003), thus facilitating accurate judgement. Due to the necessary spatial mapping capabilities of the technology (Evans et al., 2017), AR leads to a good contextualisation of the visualised content, which can support the establishing of an OM (memorising or storing information and action of recalling and remembering) (Bannon & Kuutti, 1996). This process enables the user to build up their own process knowledge by complementing human abilities and technical possibilities (Engelbart, 1988).

While the application stores the exact setting parameters and thus reveals some of the explicit knowledge of the experienced operator, the application also recognises that not all OM can be stored (Ackerman, 1998; Randall et al., 1996) by allowing the observation of the processes through the interface. This ensures that the inexperienced machine operator can build up expertise independently. Thus, AR technology provides a good transition between formal knowledge content and the local, specific, knowledges characteristic of expertise (see Chap. 4).

Due to the strongly context-dependent visualisation, AR has been found in general to lead to a lower cognitive load among users (Bresciani & Eppler, 2009). In this context, however, it was clear that some measures to reduce the cognitive load were necessary. Specifically, the results suggest that two measures could be effective in this context. Firstly, a reduction of the information to the essential process parameters is needed. Secondly, a representation in which real machine and virtual information merge as effectively as possible in one consolidated interface would help reduce that overhead.

AR seems to be particularly suitable for retrofitting machines because an interface for any machine can be created more or less independently of the hardware (Al-Maeeni et al., 2020). It establishes a direct reference to the machine by placing AR images with centimetre precision at the physical hardware itself. Through standard communication protocols, AR glasses can deliver near real-time information transmission of certain machine signals, which in turn has a positive effect on expertise sharing (see Chaps. 4 and 5).

The HoloLens assists the actual process with the visualisation of the geometry and parameters and provides a compact new interface. The evaluations show how the machine operators appreciated the ability to concentrate on only one interface. In the course of retrofitting, this means that all optical signals provided by the old, human-machine interface (HMI) must be migrated to the new interface. The separation of an HMI and an AR interface for information visualisation, according to Romero et al. (2016), should be avoided to encourage the increase of cognitive abilities by support systems. Within the AR interface, default and actual values should be placed beside each other in such a way that a direct comparison of the two values is possible. However, the results of the evaluation also show that for the near-real-time processing of the information, it is also important to display the contents selectively according to the process's progress to avoid operator misinterpretations. In concrete terms, the virtual representation of the set parameters should be built up depending on the position of the press roll.

With regard to retrofitting special machines, it is important to focus on the machine operator as a key decision-maker and to support them by providing individual sensor values. The point here is that the second-by-second skilful work practices that lead to error avoidance and smooth production require AR interfaces that can be built around the changing boundary conditions (e.g., the fluttering of the head depending on the pressure) or can allow visualising of context parameters directly in the form of process values. In addition, the AR images also allow the prominent placement and highlighting of certain real processes which particularly require the operator's expertise (e.g., marking the head with lines).

As our evaluation shows, the AR interface is a useful tool for assisting in the qualification process of inexperienced operators. The visualised information gives the operator an indication or an idea of the settings with which the head can be produced. Based on this information, the operator can gain experience and ultimately, since there is no strict target parameter specification, determine how much he can and should deviate from the given parameters (Cimini et al., 2020). The displayed default values should only serve as an orientation, and the operator must be allowed to trust their own experience. Therefore, visualisation technology must leave space for the interpretation of the values under constantly changing boundary conditions and context variables. Our application pays due respect to, and is predicated on, the concept of sociable technology Ludwig et al. (2017) and the three contexts (internal context, spatial-material context and task- and process-related context) which must be considered.

9.9.1 Limitations

Our study is, in certain respects, limited. Our application was rolled out on only one machine (a function of the need to keep as many machines working full-time as possible). We cannot say at this time what effects the technologies will have on other machines. However, from a technological point of view, further machines can be prepared for use with RetrofittAR by using ubiquitous sensor technologies. Another limitation is that the AR application is specific to the production process considered in our study. Generalising statements can only be made to a certain extent, and the results must be validated in similar settings but based on a different production process. Unfortunately, only one operator with sufficient expert knowledge was available for the eye-tracking study. Therefore, limited general conclusions can be drawn from these results, but the findings were, even so, very important for the concept and implementation (see Table 9.2). The results of the evaluation also confirm the findings of the eye-tracking study. The evaluation particularly emphasises that the interface still allows the practice of observing the head on the machine and even intensifies the practice by the fact that the operator finds the complete interface on the machine and therefore in his direct field of vision. Because only one process was considered, only some the complexities of the expertise-sharing process can be examined in detail. In addition, the implementation was only carried out for the HoloLens. Statements about other AR glasses and their suitability for use in such a scenario can therefore not be made with any great confidence. However, it is to be expected that due to the strong spatial reference of the displayed AR images (for example, the indicators for segmenting the head with the corresponding tool tips), AR glasses with no spatial mapping capabilities are unlikely to offer any added value for this use case. Nevertheless, we successfully demonstrated that, for the process considered in this paper, the AR application provided a common information space (Schmidt & Bannon, 1992) which effectively facilitated 'learning by doing'. The paper shows how configurable knowledge content can potentially be visualised utilising AR in a wider range of industrial and other contexts, thus promoting competence development. For a long-term study of expertise sharing support, an appropriation study in the company and accompanying quantitative evaluations to measure the effectiveness of systemic support are necessary.

9.10 Conclusion

The literature uses the term 'anthropocentric production systems' or 'human-centred CIM' to describe a human-oriented view of computer-assisted production processes. Such a perspective gives the essential decision-making function to the human being, largely due to the existence of various existing uncertainty factors. This is why the term 'competence-based manufacturing' is often used in this context (Brödner, 2007). In CSCW, this competence has a direct relationship to problems associated with organizational memory (OM) and the sharing of expertise and knowledge in particular (Bannon & Kuutti, 1996) and remains a vibrant issue. New technologies are said to be able to remedy this to a certain extent and to increase cognitive abilities through assistance systems (Romero et al., 2016). In this context, novel AR technologies stand out, as they can visualise highly contextualised instructions and support embedded in CPS competence building (see Chap. 5). This paper shows how AR technology can support the operator to remain at the centre of a production process and to make the essential decisions in a production process by supporting the operator in competence development (Cimini et al., 2020). Figure 9.10 shows how the prototype we present allows the operator to act as a controller in a control loop and thus remain at the centre of the production process. For this purpose, the image of a control loop is used: The target variables are defined by drawings and standards. However, deviations (fluttering head, gap, circumference) occur during production, which the operator must compensate for with the help of parameters (pressure, speed and rotation speed). RetrofittAR supports this process with the three main evaluation results shown in the figure. The spinning process carried out on the basis of the parameters is disturbed by the disturbance variables (tool abrasion and material properties), so that the head must be repeatedly checked by a quality check. AR technology is also suitable here because it allows the operator to carry out the quality check by observing the head (see Fig. 9.10) while at the same time viewing the default parameters.

The point is, as we have demonstrated, that in the real world, the risks (and costs) associated with radical change are too much for SMEs to bear. Older machinery continues to be reliable and functions well for the purposes required of it. The problems encountered are much more to do with continuity of knowledge between 'old hands' and inexperienced operators.

For the concrete answer to the research questions, the results of the study show that operators need to manage different data types, all in real time.

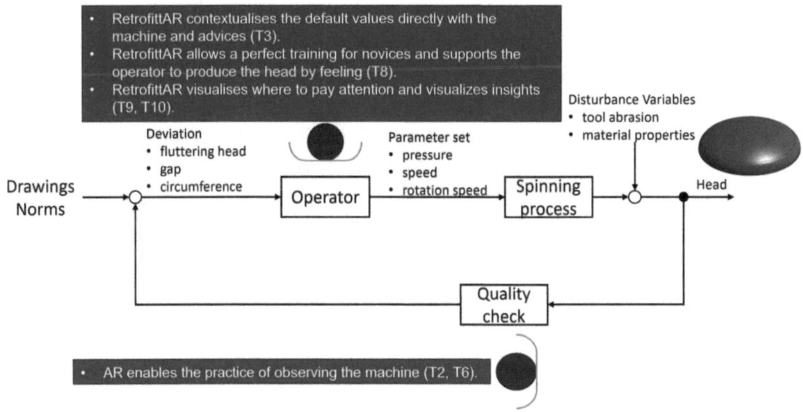

Fig. 9.10 Human machine control loop supported by RetrofittAR

- *Geometric data* represents sections of the head's final geometry to be produced and exactly positioned on the machine. A special feature is that the geometry is dynamically recalculated on the HoloLens depending on the article to be produced based on nominated parametric specifications so that no rigid computer-aided design models need to be stored. The AR display on the machine allows the machine operator to compare the actual head with the visualised geometry and compare the position of the real pressure roller with the values specified along the head's radius.
- *Default values for pressure and speed settings* are displayed along the radius of the head. Based on the experience of one machine operator, values for all heads are recorded in tabular form and displayed onto the corresponding radius sections of the real head with the help of AR glasses. These are default values which allow the successful production of a part. The defaults can be dynamically adjusted depending on the wear and tear of the tools and the raw material quality.
- *Visualisation of the live data* of the machine directly next to the default values allows a continuous target or actual comparison between the values. Thus, all the data required for the production of a head is located on the new AR interface, and the machine operator can both concentrate on the interface and observe the head on the machine directly.

Furthermore the qualitative results of the evaluations are such that we can make the following claims about the design of AR retrofitting interfaces to support expertise sharing processes.

- AR is well suited for retrofitting old machines. Retrofitting is especially effective if operators typically focus on certain areas of a machine since they can be clearly highlighted during a running production process. It is also relevant to situations where a wider field of view covering the whole of the work piece to see progress in detail during production. The interfaces can be aligned in such a way to provide a clear view of certain areas and position essential virtual content in the direct field of view.
- AR contextualises the content with the wider environment, which has a positive impact on expertise sharing processes. The contextualisation between real machine and virtual content allows inexperienced machine operators to understand previously unknown and difficult-to-explain relationships between the process and the process parameters. The results of the evaluation show how process knowledge can be gained more quickly in the course of expertise sharing on the shop floor in a real production environment—especially for old machines without numerical controls. Furthermore, the technology allows the establishment of an expertise-sharing environment in running production.
- Ensuring the operator's ability to control processes requires the interface to contain all information necessary for operation and to avoid media discontinuity. Additionally, it must also gradually build up content according to the process status and thus selectively visualise the content.

Our study clearly outlines the role that humans and their expertise play within complex production processes. The AR technologies seem to support the developing expertise of human operators where electronic controllers are not feasible (for reasons of cost etc.). Applications like RetrofittAR allow users to actively manage processes with requisite skill, even if they have little previous experience in operating the machine.

Our design case study contributes to current CSCW discourses on expertise sharing and the role of humans in manufacturing settings and provides technological interventions that illustrate how continuity of skill and knowledge can be supported in environments where neither formal training nor generic applications are adequate. Such environments, we suggest, are common.

Future studies might further investigate the appropriation of the technology, focusing on long term adaptation to the technology, its impact on workplace practice and on knowledge transfer. Subsequently, it is planned that the application will be rolled out to other use cases and machines to better assess the transferability of our findings.

Part III

Research Results

Summary 10

At the outset, the thesis suggests that the immense technological advances of recent years have also given rise to entirely new possibilities for supporting knowledge-intensive processes in the course of KES. Continuing the tradition of KES in CSCW and taking up the discussions regarding the capture, description, and demonstration of knowledge, it can be argued that novel technologies have the potential to form a new third generation of knowledge management. The thesis focuses on AR-based CPPS, due to their strong potential to support knowledge-intensive processes (Feiner et al., 1993). Nevertheless, more detailed research is needed into which processes and knowledge components of the set-up processes can be supported by which technology and specific design. A major challenge is that set-up processes are conducted on the shop floor, and these technologies must therefore fit into the working environment of experts and novices. Therefore, it is first necessary to identify and more precisely characterise processes and knowledge components of set-up processes. Building on this, the central challenges and implications for supporting the set-up processes must be identified and transferred into implementations and ultimately evaluated.

The thesis utilises a practice-based approach to answer the research questions listed at the beginning of the thesis:

- How can set-up processes be characterised?
- What are the central challenges and implications for the technical support of knowledge-intensive static and dynamic set-up processes?
- What are the requirements for an implementation to support static and dynamic set-up processes?
- How can static set-up processes be supported by AR-based CPPS, and how should holograms be designed to support them?

S. Hoffmann, *Knowledge Sharing on the Shopfloor*,
https://doi.org/10.1007/978-3-658-50133-4_10

- How can dynamic set-up processes be supported with the help of AR-based CPPS?

Regarding the first research question, Sect. 10.1.1 shows that set-up processes comprise much more than the actual assembly of tools; rather, it is a multi-stage process that includes both mechanical and non-mechanical set-up steps, which can also have a static or dynamic character. This characterisation leads to important implications for supporting set-up processes.

The second research question is summarised in Sect. 10.1.2 by focusing on embedding knowledge practices in the practice of set-up. In turn, the central challenges for the technical support of knowledge practices are derived from this. Section 10.1.2 lists the design implications for static and dynamic set-up processes.

In the following, the third research question is summarised and the implementation to support the static and dynamic set-up activities is presented. The results of research questions 1 and 2 are condensed into the requirements of the two central implementations.

The fourth research question is summarised and answered in Sect. 10.1.4 based on the main results of the evaluation. The results relate to the possibilities of supporting knowledge transfer and key findings regarding the design of the holograms are listed.

The final research question is answered in Sect. 10.1.5, which summarises the key findings regarding the support of dynamic set-up processes.

The thesis therefore makes a significant contribution to expanding the state of research at the interface between knowledge management, CPPS, and set-up processes. It explores the requirements, implications, and technical possibilities of the technologies mentioned at the level of the organisation of set-up processes regarding the subdivision into static and dynamic set-up parts, as well as at the technical support level with the help of AR-based CPPS.

10.1 Summary of Findings

The most important results of the previous chapters are highlighted in the following chapters, and they are used to provide summarized answers to the research questions.

10.1.1 Extended Characterisation of the Set-up Processes

A closer look at the set-up process reveals different phases, namely preparation (phase 1), logistic (phase 2), tool and machine set-up (phase 3), production (phase 4), inspection (phase 5), and programme iteration (phase 6). The preparation phase is characterised by the planning of the set-up sequences. Here, it is decided in which order the set-up processes are to be carried out on the machines. In the second phase, transport tasks are carried out before the machine and tools are dismantled and assembled in the third phase, after which production can begin. However, the first parts produced are usually test parts that are subjected to an inspection. Based on the inspection, adjustments are made to the machine and particularly the machine settings. The individual phases strongly differ in their characteristics and the complexity of their implementation. Therefore, a model was developed within the scope of this work, which subdivides the individual phases into static and dynamic set-up components and mechanical and non-mechanical set-up components. In addition, the individual phases are characterised regarding the possibilities of a KES process. Essentially, the thesis shows that it is possible to distinguish between relatively constant or repetitive set-up processes (static) and highly variable set-up processes (dynamic). Regarding the mechanical and non-mechanical features of the process, the results suggest that the mechanical part of the set-up process involves assembly and disassembly, while the non-mechanical part involves interaction with the machine's user interface, e.g., configuration of the CNC code. These findings represent a significant extension of the classical SMED approach by considering not only internal and external set-up parts but dynamic and static as well as mechanical and non-mechanical set-up parts (Shingo, 1985). The results of the work indicate that some of the characterisation of set-up steps can be documented. In other words, there are steps that can be easily documented due to their simple and quick explicability, whereby these steps can easily be made explicit (Nonaka et al., 2000). If this is true, then the steps are also easily systematised and proven methods from SMED are applicable (Shingo, 1985). By contrast, certain steps—especially with dynamic set-up components—can neither be explicated, documented, nor systematised. The reason for this is the high complexity of the steps due to many changing influencing factors. In these steps, the experience of the machine operator is particularly necessary (Goubergen & Lockhart, 2005). In this context, it is important to realise that there are two different set-up characteristics regarding knowledge transfer, namely static and dynamic set-up processes.

10.1.2 Central Challenges of Supporting Knowledge-Intensive Set-up Processes

Due to the complexity of the set-up process—which results from the technical conditions of the tools and machines and environmental influences such as time pressure—it is necessary to develop a technical approach adapted to the set-up process to support knowledge transfer. The technology must be able to provide a flow of information that can be consumed during a set-up process and deliver relevant, timely, and detailed information. At the same time, operators must be reliably instructed on the necessary activities to be performed. Technology should also be a source for acquiring information and as a basis for training, provided that some of the knowledge is propositional in nature. At present, this type of information is not always up to date and well maintained and not always easy to find. In this context, the possibilities of "learning by doing" (Lave & Wenger, 1991) quickly come to mind, although upon closer inspection they are limited by the fact that it is usually not a collocated workspace (Tang et al., 2007).

Due to the diversity and complexity of set-up processes, various knowledge components are involved, and the concepts listed in the KM literature do not adequately account for this complexity. Schmidt (2012) highlights that there are not only two types of knowledge but rather a 'knowledge of different things'. Besides the 'knowing how' and 'knowing that' listed by Ryle (Ryle, 1945, 2009), 'knowing what', 'knowing who', 'knowing where', and 'knowing when' are also relevant.

The question concerning how workers can share the knowledge that is embedded in their activities holds central importance. It emerges that explanations are only a secondary activity, as it is rather about demonstrating the activity. The situation in which the demonstration takes place and the recipient hold central importance. Traditional expertise-sharing approaches do not consider these boundary conditions accordingly, meaning that the essential component of embodied work is neglected. This is where the argument that new technologies can help comes in. Sensory values provide data on certain inspection points and enable the setter to determine correlations between activities and the values. A visualisation of the values adapted to the environment based on near-real-time data processing is a technical prerequisite in this respect. Given that the work is essentially conducted on tools and machines, mobile AR systems that allow the visualisation to be embedded in the working environment are suitable. There are three main design challenges to be considered at the highest level. Given that the environment of machine set-up is not—or only slightly—forgiving of errors with a constant time pressure during the execution of the set-up steps, set-up support systems have to be designed so that relevant information is provided promptly and with minimal error probability. On the one hand, codifiable information is required that can be easily recorded and passed on. However, care must be taken to ensure that the information is easily accessible and can be maintained over the life cycle of a tool or machine. On the other hand, certain tasks only seem feasible with sufficient experience, which can lead to bottlenecks in the timely execution of the set-up processes. For such cases, technical solutions should be sought that can record subtle activities of experienced employees. When recording set-up processes, it has been shown that the first-person perspective is advantageous (Fussell et al., 2003). Therefore, head-mounted applications are particularly suitable for recording set-up activities, especially since the hands are free when using such a visualisation technology and can also be recorded. An important addition is the sensory recording of deviations from target specifications. In terms of error prevention, it must be ensured that set-up processes are checked and thus take place without errors. Sensors help to monitor the work area and are thus part of the set-up support, as uncertainties are reduced and the effectiveness of technical instruction is increased through error prevention and stress reduction (Miller & Swain, 1987).

During a knowledge-transfer process, the people involved are the first focus of consideration. There are two roles, namely a knowing person and a person seeking knowledge (Nonaka & Von Krogh, 2009). Both persons meet due to an event that reveals the existence of a knowledge gap (Ackerman et al., 2013). The encounter takes place in a situation of knowledge application. In most cases, the set-up process is a real production situation. The triggering event and the gathering of people is identified as the initiating phase for a knowledge-transfer process. It is also essential to realise that the individual characteristics of the people involved influence the knowledge-transfer process. The knowledge carriers are characterised by their exact expertise and seniority. They can be internal—i.e., belonging to their own organisation—or external, i.e., belonging to a foreign organisation. Knowledge seekers are mainly characterised by the knowledge that they already possess (Sweller, 1994). As already described, the knowledge-transfer process takes place in a situation of knowledge application. The work context has a decisive influence on how the communication between the people involved in the knowledge-transfer process takes place. The exact production situation includes the production process and the machines and tools involved in the process, as well as the organisational boundary conditions. All influences taken together describe the boundary conditions under which the knowledge-transfer process takes place. In the literature, a distinction is made between an "experimental environment" for the learning process and a "practice-oriented environment" for the application of knowledge (Raelin, 1997). Due to the realisation that knowledge transfer is significantly influenced by the work context, a change in context means that what is learned in one context cannot be applied in another. A production context is particularly critical as there is little room for experimental approaches due to low error tolerances and constant time pressure. Nevertheless, according to (Raelin, 1997) experiences are an essential part of knowledge generation. In the set-up context, the experience is represented by the mastery of different situations. Feedback plays an important role in developing experiences in learning processes, since "know why" can only be achieved on the part of the knowledge seeker through feedback from a knowledgable person (Banker et al., 1993; Stone & Stone, 1984). The intensity of feedback correlates with the complexity of the task (Wollersheim et al., 2015). New technologies have an important role to play in providing feedback. Through the targeted use of support systems—as presented in the knowledge-transfer model—fewer human resources are needed and the possibilities of integrating feedback into the workplace ensure that the concrete production context is always referenced. In addition to the aforementioned work context, it is also desirable that the entire organisational context is open to KES processes. In order to achieve this openness, the technology supporting KES must be able to ensure the reliability of information and protect private knowledge. The fact that the work context can also be a hindrance is made clear by the fact that external learning environments are preferred for learning to operate new machines in some cases. These environments can be advantageous because there are minimal time restrictions.

The knowledge flow takes place between the knowledge seeker and the knowledgeable person. Usually, the initiation of the knowledge flow takes place through the knowledge seeker, either a direct exchange or an indirect exchange via storage media takes place. If technology is used, it must also be able to transfer embodied knowledge. This is where CPS technologies come in, as the communication between knowledge seeker and knowledgeable person can be designed in a new way. The technical possibilities to facilitate the collection of the required knowledge and present knowledge components in different formats are crucial, e.g., in the form of text representations and static or dynamic graphical representations such as images, videos, and AR overlays. As suggested in the literature, the possibility of different presentation formats is elemental to the success of sharing, especially in technical environments (Patterson et al., 2014; Wright, 1977). However, the results of the work also show that creating storable documentation and instructions is challenging for various reasons.

(i) *Identification of Different Design Implications for Static Set-up Processes*

The challenges make it clear that conventional technologies in the form of tablets and smartphones can only be integrated into everyday work and knowledge-transfer processes to a limited extent. The potentials of possible new technologies are also discussed, formulated through concrete design implications.

The implications are guided by the CPS architecture model established by Lee et al. (2015). The first step is to capture contextual information related to the set-up process, which the authors refer to as the "smart connection level". It is about recording sensors to capture the set-up processes—in addition to the knowledge-based process steps—and using them as a basis for further steps and optimisations. Two main areas stand out for sensor-based recording: on the one hand, there are the logistical activities, where sensing systems offer possibilities to record and visualise the distances travelled, which enables working out further decisions and optimisations of the set-up steps; and on the other hand, sensory recording of the tools and the tool positions on the machine play an important role. In this way, the quality of the set-up process can be ensured by sensors based on positions and dimensions. At the same time, these data serve as a source of feedback for the worker who carried out the set-up process (Vicente, 2000).

In addition to the real data from the set-up process, purely virtual data is also an essential component of set-up support. Here, again a comparison can be drawn with Lee et al. (2015). The second level is called the conversion level, which is concerned with converting data into information. The virtual data of the set-up process provides information in the form of virtual instructions. First, virtual simulation data is used to avoid collisions. The data originates from a virtual design environment, which enables carrying out not only kinematic simulations but also mechanical correlations and even deriving machine settings (Altintas et al., 2005; Jönsson et al., 2005). In addition to the simulated data, AR environments can be used to record and visualise information on the exact position of a tool on the machine and the movement sequence of the assembly. The basis for the AR representation is CAD data of the machine and tool components.

Thus, the data basis comprises sensory and virtual data and information, which must be combined at the next level to form a set-up support. Since set-up processes are mostly manual processes, a human-centred aggregation of the data must occur. If we look again at the CPPS architecture of Lee et al. (2015) in this context, we find the cyber level with the twin model as a representation of machines and components at the third level. In the course of supporting the set-up process, the data of the two lower levels are combined at this level. In the process, the sensory data and the AR instructions created are combined into set-up instructions in an authoring environment. In order for the data to represent added value in the course of knowledge transfer, it must be prepared and presented in a situationally relevant, detailed, and problem-oriented manner (Lee et al., 2015). Specifically, a complete workflow for creating the set-up instructions comprising sensory data and holograms in combination with videos, photos, and short text information must be processed.

After the data aggregation, a knowledge transfer must be enabled. For this purpose, the data and instructions for the corresponding set-up interactions are presented in a replay mode. In addition, the complexity must be reduced by only presenting context-sensitive data containing essential key elements of a set-up process (Schuh et al., 2015). Given that mechanical set-up processes often involve embodied action parts, instructions that represent the set-up interactions play a central role. Set-up interactions that have been recorded must be presented in a way that is adapted to the environment. It is important that the instructions are available independently of the presence of knowledge carriers to maintain the knowledge transfer. The entire set-up instruction usually has several steps represented with holograms, photos, videos, and short texts, and must be processed in a certain order.

However, ultimately it is not sufficient during mechanical set-up activities to display and reproduce instructions that have been generated once; rather, the fact that the set-up process is optimised through experience over time must be taken into account. Therefore, instructions must also be adapted over time (Vicente, 2000). Lee et al. (2015) also list a "configuration layer" in the CPPS architecture, whereby it must be possible to optimise the created instructions based on experience and sensory values and subsequently make them available again.

(ii) *Identification of Different Design Implications for Dynamic Set-up Processes*

Dynamic set-up parts are characterised by the interaction of the operator with the control/operator interface of the machine. It usually involves the operator making adjustments to the machine to prevent producing defective products.

An essential requirement is to make the machine operations visible while the machine is in operation. Accordingly, instructions must always be precisely adapted to the machine. If this requires data from the machine, older machines must be extended so that the machines provide digital values. However, the instructions on the machine must be presented in such a way that they integrate perfectly into the process and are very easy to follow when operating the machine. A prerequisite for this is that there is only one interface for all necessary process data and information.

It has been shown that the operator often has to make essential decisions based on recorded quality data from production. Since collecting quality data often involves observations of the machine and the workpieces, instructions should be perfectly integrated into this observation practice by placing them in the operator's direct field of vision and providing only one interface for all necessary data and information. Input data and machine parameters should also be directly related, allowing the operator to gain a deeper understanding of the relationship between input parameters, machine parameters, and quality outcome. Another implication is that the input parameters have a strict location reference to the machine or the workpiece, meaning that the position of the machine or the workpiece must be known. Consequently, the geometries of the workpiece must be known, or the positions of the machine axes must be transmitted digitally. If different workpieces are produced, it is necessary to consider the different geometries of the workpieces. Summarising these implications, AR technology is particularly suitable for visualising instructions as the technology can place the holograms in a location-specific manner due to the recognition of the environment (Evans et al., 2017).

Finally, disturbance variables play a relevant role in the execution of a production process and often lead to quality deviations, if not taken into account. Therefore, process-specific disturbance variables must also be considered and the instructions must be adapted accordingly.

As with the implications for the static set-up process, it is also true for the dynamic set-up process that a knowledge-transfer process must be realised independently of the presence of people. However, in the dynamic set-up process, the focus is not so much on knowledge transfer between people mediated by technology but rather the representation of input parameters, machine parameters, and quality results.

10.1.3 Implementation of Static and Dynamic Set-up Support

Two AR-based set-up supports were implemented based on the design implications listed. The static set-up support is a CPPS system comprising several sensors and visualisations. The database comprises CAD data of the machine and the tools. The CAD models already contain the final assembly positions of the tools in relation to the machine and the other tool components. CAD models of the machines are important when machine components are part of the set-up process. However, the CAD data must be prepared in Unity for display on the HoloLens.

Both the number of vertices of the model and the assembly structure usually have to be adjusted. Changing the CAD model means that the model has to be reprocessed in Unity and the application has to be recompiled on the HoloLens. In addition, a sensory component comprising a 3D camera with corresponding camera software and from the sensor data of the HoloLens was implemented. The 3D camera monitors the position and dimensions of the tools and the HoloLens provides path data regarding the set-up paths. All data is processed in the set-up tool. From the CAD data, a step-by-step set-up instruction comprising animated AR holograms, videos, photos, and short texts is created and saved in an authoring environment. These instructions can be consumed in playback mode without further editing of the set-up instruction. After appropriate configuration, the sensor data of the camera provide indications concerning whether the correct tool has been mounted on the appropriate position on the machine axis. The visualisation of the result of the camera check is also displayed on the HoloLens as a coloured marking of the holograms. The instructions created with the HoloLens can be exported to a tablet application and subjected to a check there. Based on the set-up performance—comprising times and distances recorded with the HoloLens—optimisations of the set-up sequence can be elaborated, implemented, saved, and exported to the HoloLens again.

Based on the knowledge that dynamic set-up processes fundamentally differ from static processes, a second application was developed to support dynamic set-up processes. This is an application that is very strongly adapted to the process to be supported because here an integration of process parameters with geometric dimensions takes place, and thus an extreme adaptation to the subtleties of the processes is required.

Essentially, the application comprises three essential data elements, calculated on a HoloLens, stored, or transferred to it via a WebSocket connection. First, the geometry of the workpiece to be produced is calculated using a few parameters and formulas defined in standards. The calculated geometry serves as an anchor point for representing the target parameters for speed and pressure. These were universally recorded in tables and displayed depending on the workpiece to be produced and the machine used. The third data component is the machine data in the form of a signal light of a pressure and a speed value. The data is displayed next to the target parameters for speed and pressure. Overall, the result is a compressed AR interface that displays all this data in a compressed form and thus considers the complex requirements of integration into everyday operations.

10.1.4 Support for Static Set-up Activities

The support of static set-up activities with the help of AR-based CPPS set-up support shows five topics.

- First, the support relieves the experts' and novices' time (Nilsson & Johansson, 2008) as it is available independently of the person after an initial recording of the instruction, and thus there are no strict time commitments on either side.
- The instructions must fit into the work context in a special way, in the sense that they can be consumed in parallel with the execution of the activities. It is therefore important that they are simple, easy to understand and as embedded as possible in the work environment.
- AR instructions have the potential to establish a standard (Shingo, 1985) to a process, as all people involved have the same basis. In addition, an autonomous learning environment (Paas et al., 2003) is established. Not only are novices provided with essential instructions, but experienced operators are also actively reminded of rarely performed set-up processes. At the same time, there are doubts of the experienced operators about the learning progress that can be achieved through the instructions. There is a risk that the instructions will lead to reliance on them and their constant availability.
- The representation of the instruction using animated holograms was not always perceived as sufficiently accurate, while the holograms only depict the rough set-up interactions (Bloodgood & Salisbury, 2001) but do not show every move in detail. Therefore, the supplementary videos were found to be very helpful. The videos are a useful complement to the holograms and support building an organisation-wide repository of tacit and explicit knowledge (Bloodgood & Salisbury, 2001).
- During authoring, the holograms based on CAD data and the animations were perceived as positive. However, the experts were unable to adequately represent all details. Especially the contact points of machine and tool need to be represented more precisely. In addition, created instructions have to be adapted because instructions change over time and the expert discovers errors or potential for improvement when recording.

The results show that holograms in an AR environment can be helpful for set-up processes. The representation of the tool components is based on CAD models. This is obvious in this use case since CAD models of the tools or machine components already exist in most cases. However, the existence of CAD models cannot be taken for granted because no CAD models exist in other use cases. If CAD models have to be created first in such cases, it is almost impossible for HoloLens users since the software for modeling is not available and the expertise is mostly lacking. Therefore, the question concerning whether there are alternative forms of representation in extension or in addition to holograms based on CAD models is justified. Particular attention is paid to representation and hardware in supporting knowledge-intensive activities (Nakanishi et al., 2010). Therefore, considering different ways of visualising holograms on the same hardware holds particular interest. For this purpose, very simple holograms comprising arrows and wire mesh boxes are compared with holograms based on 3D models. In addition, the work shows the relevance of supplementary videos. The target variables examined in this thesis are the completion time, error rate, and cognitive load. Overall, it can be summarised that the type of hologram has no significant influence on the completion time, aligning with other studies (Funk et al., 2015). There are significant differences in the number of errors, whereby the holograms based on 3D models perform significantly better than those based on abstract shapes (Blattgerste et al., 2017). Moreover, it is clear that video leads to a significant reduction in the number of errors. However, concrete holograms also have disadvantages as it has been found that they lead to a higher cognitive load. Finally, machine operators often have no knowledge regarding the creation of AR content (Gattullo et al., 2019), prompting the conclusion that abstract holograms offer a good compromise for the representation of instructions when CAD models are not available. Otherwise, representations based on CAD models are preferable due to the lower error rates. In any case, videos are generally a good complement when complex activities are involved and the holograms do not represent details.

10.1.5 Support for Dynamic Set-up Activities

Essentially, the support provided by the HoloLens application for static set-up processes was mainly considered helpful by inexperienced machine operators. By contrast, experienced machine operators perceive the displayed information as redundant, leading to an overload of information (Kalyuga et al., 2003). This is supported by the findings from dynamic set-up processes, although the representation of the holograms fundamentally differs between static and dynamic set-up processes. As part of the dynamic set-up process, the visual representation of the target parameters to be set and the parallel possibility of observing the production process are to be emphasised. Ultimately, this creates an essential basis for the novice to carry out the process independently and gather experiential knowledge on this basis. The added value of AR support often lies in adapting the holograms to real objects, ensuring that the visualisation adapts to the current context and task (Bannon & Kuutti, 1996). However, especially with AR instructions, creating the content is not trivial as it involves three-dimensional objects (Nebeling & Speicher, 2018). Therefore, a compromise is often made by combining standard geometries into a representation that is adapted to the process but still relatively abstract (Radkowski et al., 2015). Calculating contours and volumes based on defined geometric dependencies is a useful idea to simplify the creation of 3D geometries. However, this is only possible for products defined by norms and standards. In the production of heads, the standardisation of geometries is very advanced, and thus a virtual head generator was implemented on the HoloLens to calculate the AR geometries based on a few article data.

The results also show that AR technologies are particularly suitable for supporting knowledge transfer in the case of rapidly changing knowledge content. It has been shown that the combination of the representation of default values and the simultaneous observation of the real process and the behaviour of the workpiece leads to a very good contextualisation of the visualised values and ultimately results in developing an OM (Bannon & Kuutti, 1996) and a process knowledge (Engelbart, 1988). Furthermore, it has been shown that the default values provide a basis of explicit knowledge but still leave sufficient room for building up own experiences. This counteracts the danger of novices relying only on the default values and thus preventing the development of their own process knowledge. In addition to the effect of AR visualisations on the cognitive load (Bresciani & Eppler, 2009), further helpful measures in the form of a reduction to the essential default values and the compilation of all essential data for the process on a machine-oriented AR interface have been shown to reduce the cognitive load. The comparison of target and actual values enables a direct comparison of both values. In addition, selective representations of values should be aimed for, in the sense that only the values for the current position are displayed according to the position of the machine, whereas all other values are hidden.

Another finding is that AR technology is suitable for retrofitting machines, as new interfaces for the machines can be built independently of the machine hardware (Al-Maeeni et al., 2020). Finally, it must be mentioned that AR technology is an assistance system that sustainably strengthens the role of humans in a production process. When instructions are presented, it is ensured that humans can gain their own experience and expertise by providing orientation to the specifications and leaving sufficient room for interpretation and adaptation. Therefore, the operator is supported in competence development (Cimini et al.,

2020), which the literature refers to as 'competence-based manufacturing' (Brödner, 2007). Ultimately, the human being takes on the role of a controller in a control loop comprising specifications, machines, parameters, and quality results. This means that people are still the brains of the process but AR technologies support developing an OM and help people to take on the role of a process controller.

10.2 Conclusion

The thesis has examined set-up processes on industrial machines against the background of technology-supported knowledge-transfer processes. Based on empirical studies in companies, the work shows that set-up processes are highly complex and that only considerations to support mechanical activities fall far short of the mark. Non-mechanical activities are also very important and represent a major challenge in operational practice. The thesis takes up essential results of the literature regarding knowledge and expertise sharing and its technical support, and extends the state of the art by demonstrating the support of KES practices using AR-based CPPS technology based on set-up processes. These technologies have the potential to form a new generation of support for KES processes.

The work does not represent a generally valid solution for the support of set-up processes but shows in detail the possibilities and limitations of AR-based CPPS. In particular, the results show that the support of dynamic set-up processes must be highly customised to the process. It also becomes clear that the technologies can neither replace expertise nor mechanical skills but rather represent an opportunity for knowledge seekers to gain a better insight into the process and thus enable them to build up their own expertise in the form of process knowledge.

To summarise the work, the following implications for designs can be derived:

- It is recommended to first derive the characteristics of the set-up process to be supported to then show the exact knowledge-transfer process in a second step. Based on this work, set-up processes can be broken down into six successive steps and two key distinctions can be identified between dynamic and static set-up steps. Static set-up processes have a repetitive character, and dynamic set-up processes are adapted to different environmental parameters each time they are carried out. It is therefore sufficient to create static AR instructions for static set-up processes, whereas dynamic AR instructions must be used for dynamic set-up processes.

- Knowledge transfer in companies is ensured by a knowledgeable person and a knowledge seeker meeting in a situational context and exchanging information and data either directly or mediated by technology. The two people can be described by key characteristics. The knowledgeable person is characterised by a high proportion of embodied knowledge, the presence of time constraints, and a certain degree of knowledge ownership, while the knowledge seeker is characterised by a learning pace, a degree of expertise, and a certain level of dependency. Technological AR support for knowledge transfer must first enable the expert to create AR instructions. The content should also be able to capture embodied knowledge aspects and the creation must not generate any significant overhead. Consequently, the content creation must be embedded in the working practice of the knowledgeable person. On the knowledge seeker's side, the reproduction of AR content should correspond to the learning pace and the degree of expertise. These aspects show that the situational context and the characteristics of the people must be taken into account when designing the technologies. Knowledge transfer models should provide a description of the situational contexts and personal characteristics.
- AR instructions for set-up processes must utilise the advantages of AR technology. Specifically, this means that the instructions must always be created relative to a machine position or a tool position. The tracking technology of the hardware is essentially able to place the holograms in relation to the location. However, the disadvantages of tracking technologies are also evident in practical applications. During set-up processes, the tools might have to be positioned with an accuracy of less than 1 cm, which current tracking technologies are incapable of achieving. It must therefore first be ensured that the application is suitable for support with AR technology.
- In principle, AR instructions are instructions for carrying out certain manual operations on a machine or tool. Given that errors or omissions by humans can occur when executing the instructions, it is advantageous if the manual steps are checked by sensory detection of the workspace. The work has shown that the tool size and tool position are useful data to be recorded by sensors.
- Holograms should be designed in such a way that they do not overlay the real world but rather blend in with the real environment. Holograms should therefore be designed in such a way that they display location-based set-up information with minimised hologram appearance. This is particularly important for set-up processes as the assembly and disassembly positions must always be clearly visible to the setter. The animation of the set-up interaction, representation of the support surfaces, and representation of the attachment points should highlight holograms so that the most comprehensive

set-up information possible is still transmitted with a minimum appearance of holograms.
- A distinction can be made between abstract and concrete representations of holograms. Abstract holograms are easier and quicker to create but do not show all the details of a set-up step. Concrete holograms are more complex to create as they have to be derived from a CAD data set, although they show more details. Abstract representations are only sufficiently accurate for set-up processes in combination with videos.
- For dynamic set-up activities, the instructions must be structured in such a way that the machine operator is kept in the human-machine control loop and the contextualisation between the real machine and virtual content promotes an understanding of difficult-to-explain relationships between the process and the process parameters.
- Dynamic set-up settings require displaying different set-up data compared to static set-up processes. In addition to variable geometric data of the product to be produced, default setting parameters of the machine for manufacturing the products and current live data from the machine must be integrated into the AR holograms. In order to avoid interrupting the important observation practice of the machine operator's manufacturing process, the holograms must contain all process-relevant representations to prevent media discontinuity.
- AR holograms can also be used to realise complete information interfaces for old machines. AR technology can therefore retrofit old machines economically without having to modernise the machines themselves.

10.3 Outlook

The presented results and contents of the work provide essential insights into the use of modern AR-based CPPS in knowledge-intensive set-up processes. Due to the immense diversity of set-up processes and the resulting specific requirements for KES support, the results of this work should be further evaluated and subjected to long-term studies. The following seven main areas can be identified for further research into AR-supported knowledge management practices.

- *Cooperative aspects of set-up:* The work focusses on set-up steps conducted by one setter. In the environment of large machines, several setters are entrusted with the task of setting up and carrying out the job in a cooperative manner. Therefore, further steps in researching the use of AR in manual processes should also examine the cooperative aspects of the activity and its support.

Two forms are conceivable here: first, a defined division of the set-up steps among several setters, whereas each setter takes on a single step of an instruction and carries it out; and second, several setters carry out a step together. In both cases, there are new requirements for the technology and design of the instructions, which entail further research activities. In addition, when researching the cooperative dimensions, the situational contexts and personal characteristics must be re-evaluated and, at best, presented in a knowledge transfer model based on this work.

- *Authoring AR instructions:* The creation of AR instructions is an essential component in the course of knowledge transfer. The work shows possibilities for authoring in the context of static set-up processes, although the situation is much more difficult for dynamic set-up processes (see Chaps. 7and 9). Due to the constantly changing content, it is only logical that new ways of creating flexible instructions must be found. Some approaches are listed in the thesis, such as the creation of product geometries directly in the AR application with the aid of geometric relationships. If no geometric relationships can be derived, it is necessary to research new ways of creating geometries. From a knowledge-transfer perspective, it is important that the creation takes place on the shop floor; otherwise, the flow of knowledge between the knowledge seeker and knowledgeable person might be interrupted.
- *Further development of AR hardware:* The further development of AR hardware also points to fundamental changes in the visualisation of holograms and HMDs. On the one hand, it can be observed that the development of AR hardware and software frameworks has fallen somewhat out of the focus of hardware manufacturers. On the other hand, general technological developments also promise major potential for optimising the complex hardware; for example, through better sensors, and eliminating the mentioned shortcomings of tracking technology Chap. 7. There is also a need for further optimisation of the ergonomic aspects of the HMD. Further hardware developments in the area of sensory detection of the environment and display technologies are necessary to increase user acceptance of the technology.
- *Validation of the new generation of knowledge and expertise sharing:* The thesis shows that new technologies are creating new opportunities for KES, particularly in the area of embodied knowledge, and that this is initiating a possible new generation of KES (see Chap. 5). Subsequent work in the tradition of researching KES processes must show whether AR-based CPPS have truly shaped a new generation of KES. Ultimately, it is only possible to judge whether the technology has given rise to a new generation retrospectively. This can only be taken further if this type of support for KES is transferred

to other application areas, such as its use in the crafts or healthcare sector is conceivable.

- *Combination of AR and artificial intelligence:* In addition, research must be conducted into the role of AI—and embodied AI in particular—in supporting set-up processes in combination with AR. Due to the aforementioned need for research in the area of authoring, further work should consider how authoring content can be created with AI. Specifically, it is about the possibilities of designing holograms of the tools, machines and workpieces to be set up on the basis of CAD data and scanned objects. The work shows that concrete representations of the machines and tools provide added value for the KES, but are not always available or do not have the necessary level of detail (see Chap. 8). This gap had to be closed in the course of further research. In this context, it is also necessary to investigate how modern scanning technologies (e.g. LiDAR) can offer further added value in the creation of holograms.
- *Contribution to the dimensions of sustainability*: Overall, the focus of the work is placed on supporting KES and thus strengthening a company's human capital. As a result, the work also contributes to the social dimension of sustainability, while aspects of optimising set-up processes and their support through AR are also addressed. In combination with improved knowledge management, the work thus also provides approaches for an economic dimension of sustainability. From today's perspective, it is also appropriate to consider set-up processes from the ecological dimension of sustainability. For this purpose, knowledge management practices must be linked to the life cycles of products, machines, and tools. In this context, the possibilities of CPS need to be explored in more detail. The thesis demonstrates that AR and CPS technologies complement each other, for example by detecting the position of tools using sensors (see Chap. 7). The data from the tools and machines is highly relevant for the development of set-up process life cycles.
- *Research into dynamic set-up processes:* The work presents research results relating to an exemplary dynamic set-up process on a machine for the production of heads (see Chap. 9). The complexity of the parameter setting of machines is by no means limited to the listed application. The challenges in the course of dynamic set-up processes are extremely large and are further intensified by demographic change. An important approach here is also to consider the life cycles of tools, machines and products. A key result of the study presented in Chap. 9 is that the setting parameters for the production process must be recorded and visualised depending on the article, the machine and the tool. This makes a life cycle analysis of the article, machine and tool unavoidable. The possibilities of CPS technologies for recording life

cycles must therefore be further researched. Based on the process described in Chap. 9, the wear values of the tools and the material characteristics of the raw material must be set in relation to the setting parameters of the machine. Interdisciplinary research approaches are required to investigate the relationships and correlations of dynamic set-up processes.

This work contributes to supporting people in carrying out knowledge-intensive set-up activities. The points listed in this chapter indicate further research trajectories in this tradition. Advanced technologies and their research in the field of CSCW and HCI should continue to contribute to the augmented workforce and prompt transformations through empowering people.

References

Ackerman, M. S. (1998). Augmenting the organizational memory: A field study of Answer Garden. *ACM Transactions on Information Systems, 16*(3), 203–224. https://doi.org/10.1145/290159.290160

Ackerman, M. S., Dachtera, J., Pipek, V., & Wulf, V. (2013). Sharing knowledge and expertise: The CSCW view of knowledge management. *Computer Supported Cooperative Work: CSCW: An International Journal, 22*(4–6), 531–573. https://doi.org/10.1007/s10606-013-9192-8

Ackerman, M. S., & Halverson, C. (1999, January). Organizational memory: Processes, boundary objects, and trajectories. *Proceedings of the 32nd Annual Hawaii International Conference on Systems Sciences.* https://doi.org/10.1109/HICSS.1999.772789

Ackerman, M. S., & Halverson, C. (2004). Sharing expertise: The next step for knowledge management. In M. Huysman & V. Wulf (Eds.), *Social Capital and Information Technology* (pp. 273–299). MIT Press.

Ackerman, M. S., & Malone, T. W. (1990). Answer Garden: A tool for growing organizational memory. *Proceedings of the ACM SIGOIS and IEEE CS TC-OA Conference on Office Information Systems (COCS '90),* 31–39. https://doi.org/10.1145/91478.91485

Ackerman, M. S., & McDonald, D. W. (1996). Answer Garden 2: Merging organizational memory with collaborative help. *Proceedings of the 1996 ACM Conference on Computer Supported Cooperative Work (CSCW '96),* 97–105. https://doi.org/10.1145/240080.240203

Ackerman, M. S., Pipek, V., & Wulf, V. (2003). *Sharing expertise: Beyond knowledge management* (M. S. Ackerman, V. Pipek, & V. Wulf, Eds.). MIT Press.

Ackermann, B., Schorta, S., Krancher, O., North, K., & Schildknecht, K. (2018). Der Vorgesetzte als Initiator und Motor des Wissenstransfers. In *Erfolgreicher Wissenstransfer in agilen Organisationen: Hintergrund—Methodik—Fallbeispiele.* https://doi.org/10.1007/978-3-658-19468-0_7

Adamic, L. A., Zhang, J., Bakshy, E., & Ackerman, M. S. (2008). Knowledge sharing and Yahoo answers. *Proceedings of the 17th International Conference on World Wide Web—WWW '08,* 665–674. https://doi.org/10.1145/1367497.1367587

Agrawala, M., Phan, D., Heiser, J., Haymaker, J., Klingner, J., Hanrahan, P., & Tversky, B. (2003). Designing effective step-by-step assembly instructions. *ACM SIGGRAPH 2003 Papers—SIGGRAPH '03,* 828–837. https://doi.org/10.1145/1201775.882352

S. Hoffmann, *Knowledge Sharing on the Shopfloor*,
https://doi.org/10.1007/978-3-658-50133-4

Agresti, A., & Finlay, B. (2008). *Statistical methods for the social sciences* (4th ed.). Pearson/ Prentice Hall.

Akçayır, M., & Akçayır, G. (2017). Advantages and challenges associated with augmented reality for education: A systematic review of the literature. *Educational Research Review, 20,* 1–11. https://doi.org/10.1016/j.edurev.2016.11.002

Alexander, C. (1977). *A pattern language: Towns, buildings, construction.* Oxford University Press.

Al-Maeeni, S., Kuhnhen, C., Engel, B., & Schiller, M. (2020). Smart retrofitting of machine tools in the context of industry 4.0. *Procedia CIRP, 88*(June), 369–374. https://doi.org/ 10.1016/j.procir.2020.05.064

Altintas, Y., Brecher, C., Weck, M., & Witt, S. (2005). Virtual machine tool. *CIRP Annals, 54*(2), 115–138. https://doi.org/10.1016/S0007-8506(07)60022-5

Anderson-Levitt, K. M. (2006). Ethnography. In *Handbook of complementary methods in education research.* Routledge.

Argote, L., & Ingram, P. (2000). Knowledge transfer: A basis for competitive advantage in firms. *Organizational Behavior and Human Decision Processes, 82*(1), 150–169. https:// doi.org/10.1006/obhd.2000.2893

Argote, L., Ingram, P., Levine, J. M., & Moreland, R. L. (2000). Knowledge transfer in organizations: Learning from the experience of others. *Organizational Behavior and Human Decision Processes, 82*(1), 1–8. https://doi.org/10.1006/obhd.2000.2883

Argyris, C. (2021). Tacit knowledge and management. In R. J. Sternberg & J. A. Horvath (Eds.), *Tacit knowledge in professional practice* (pp. 137–154). Lawrence Erlbaum Associates Publishers. https://doi.org/10.4324/9781410603098-15

Arth, C., Grasset, R., Gruber, L., Langlotz, T., Mulloni, A., & Wagner, D. (2015). The history of mobile augmented reality. *arXiv.* http://arxiv.org/abs/1505.01319

Aschenbrenner, D., Li, M., Dukalski, R., Casper Verlinden, J., & Lukosch, S. (2018). Exploration of different augmented reality visualizations for enhancing situation awareness for remote factory planning assistance. *IEEE VR, April,* 3–7. https://doi.org/10.13140/RG.2. 2.14819.66083

Ayres, L. (2016). Thematic coding and analysis. In L. M. Given (Ed.), *The Sage encyclopedia of qualitative research methods* (Vols. 1 & 2, pp. 868–869). SAGE Publications, Inc. https://doi.org/10.4135/9781412963909

Azuma, R. T. (1997). A survey of augmented reality. *Presence: Teleoperators and Virtual Environments, 6*(4), 355–385. https://doi.org/10.1162/pres.1997.6.4.355

Badham, R. (1991). Human-centred CIM. Informating the design/manufacturing interface. *Futures, 23*(10), 1047–1060. https://doi.org/10.1016/0016-3287(91)90071-9

Baheti, R., & Gill, H. (2011). Cyber-physical systems: From theory to practice. In *The impact of control technology* (Issue 1). https://doi.org/10.1145/1795194.1795205

Baird, K. M., & Barfield, W. (1999). Evaluating the effectiveness of augmented reality displays for a manual assembly task. *Virtual Reality, 4*(4), 250–259. https://doi.org/10.1007/ BF01421808

Bangor, A., Kortum, P. T., & Miller, J. T. (2008). An empirical evaluation of the system usability scale. *International Journal of Human-Computer Interaction, 24*(6), 574–594. https://doi.org/10.1080/10447310802205776

Banker, R., Potter, G., & Schroeder, R. G. (1993). Reporting manufacturing performance measures to workers: An empirical study. *Journal of Management Accounting Research, 5.*

Bannon, L. J. (1995). From human factors to human actors: The role of psychology and human-computer interaction studies in system design. In R. M. Baecker, J. Grudin, W. A. S. Buxton, & S. Greenberg (Eds.), *Readings in human–computer interaction* (2nd ed., pp. 205–214). Kaufmann Morgann. https://doi.org/10.1016/B978-0-08-051574-8.50024-8

Bannon, L. J., & Kuutti, K. (1996). Shifting perspectives on organizational memory: From storage to active remembering. *Proceedings of the 29th Hawaii International Conference on System Sciences (HICSS'96), 3,* 156–167. https://doi.org/10.1109/HICSS.1996.493187

Bardram, J., & Bossen, C. (2005). Mobility work: The spatial dimension of collaboration at a hospital. *Computer Supported Cooperative Work: CSCW: An International Journal, 14*(2), 131–160. https://doi.org/10.1007/s10606-005-0989-y

Baumard, P. (1999). *Tacit knowledge in organizations.* SAGE Publications.

Beach, W. A., & LeBaron, C. D. (2002). Body disclosures: Attending to personal problems and reported sexual abuse during a medical encounter. *Journal of Communication, 52*(3), 617–640. https://doi.org/10.1093/joc/52.3.617

Benford, S., Bowers, J., Fahlén, L. E., Greenhalgh, C., & Snowdon, D. (1995). User embodiment in collaborative virtual environments. *Proceedings of the International Conference on Human Factors in Computing Systems (CHI'95),* 242–249. https://doi.org/10.1145/223904.223935

Benford, S., Mariani, J., Navarro, L., Prinz, W., & Rodden, T. (1993). MOCCA: An environment for CSCW applications. *COCS '93: Proceedings of the Conference on Organizational Computing Systems,* 172–177. https://doi.org/10.1145/168555.168573

Benyon, D. (2010). *Designing interactive systems: A comprehensive guide to HCI and interaction design* (2nd ed.). Addison Wesley.

Bhattacharya, B., & Winer, E. H. (2019). Augmented reality via expert demonstration authoring (AREDA). *Computers in Industry, 105,* 61–79. https://doi.org/10.1016/j.compind.2018.04.021

Blandford, A., Cox, A., & Cairns, P. (2008). Controlled experiments. In *Research methods for human-computer interaction* (pp. 1–30). CUP.

Blattgerste, J., Renner, P., Strenge, B., & Pfeiffer, T. (2018). In-situ instructions exceed side-by-side instructions in augmented reality assisted assembly. *Proceedings of the 11th Pervasive Technologies Related to Assistive Environments Conference,* 133–140. https://doi.org/10.1145/3197768.3197778

Blattgerste, J., Strenge, B., Renner, P., Pfeiffer, T., & Essig, K. (2017). Comparing conventional and augmented reality instructions for manual assembly tasks. *Proceedings of the 10th International Conference on Pervasive Technologies Related to Assistive Environments,* 75–82. https://doi.org/10.1145/3056540.3056547

Bloodgood, J. M., & Salisbury, W. D. (2001). Understanding the influence of organizational change strategies on information technology and knowledge management strategies. *Decision Support Systems, 31*(1), 55–69. https://doi.org/10.1016/S0167-9236(00)00119-6

Bødker, S. (2015). Third-wave HCI, 10 years later—Participation and sharing. *Interactions, 22*(5), 24–31. https://doi.org/10.1145/2804405

Bolisani, E., & Scarso, E. (1999). Information technology management: A knowledge-based perspective. *Technovation, 19*(4), 209–217. https://doi.org/10.1016/S0166-4972(98)00109-6

Bossen, C., Christensen, L., Grönvall, E., & Vestergaard, L. S. (2013). CareCoor: Augmenting the coordination of cooperative home care work. *International Journal of Medical Informatics, 82*(5), e189–e199. https://doi.org/10.1016/j.ijmedinf.2012.12.005

Bottani, E., & Vignali, G. (2019). Augmented reality technology in the manufacturing industry: A review of the last decade. *IISE Transactions, 51*(3), 284–310. https://doi.org/10.1080/24725854.2018.1493244

Bowers, J., Button, G., & Sharrock, W. (1995). Workflow from within and without: Technology and cooperative work on the print industry shopfloor. In H. Marmolin, Y. Sundblad, & K. Schmidt (Eds.), *Proceedings of the Fourth European Conference on Computer-Supported Cooperative Work* (pp. 51–66). Springer. https://doi.org/10.1007/978-94-011-0349-7_4

Bowers, J., O'Brien, J., & Pycock, J. (1996). Practically accomplishing immersion: Cooperation in and for virtual environments. *Computer Supported Cooperative Work,* 5(4), 327–358. https://doi.org/10.1007/BF00133004

Braun, V., & Clarke, V. (2012). Thematic analysis. In H. Cooper (Ed.), *APA handbook of research methods in psychology, Vol 2: Research designs: Quantitative, qualitative, neuropsychological, and biological* (pp. 57–71). American Psychological Association. https://doi.org/10.1037/13620-004

Bresciani, S., & Eppler, M. (2009). The risks of visualization: A classification of disadvantages associated with graphic representations of information. *Information Visualization, 8*(3), 144–156. https://doi.org/10.1057/ivs.2009.29

Brödner, P. (2007). From Taylorism to competence-based production. *AI & Society, 21*(4), 497–514. https://doi.org/10.1007/s00146-007-0087-4

Brown, J. S., & Duguid, P. (1994). Borderline issues: Social and material aspects of design. *Human-Computer Interaction, 9*(1), 3–36. https://doi.org/10.1207/s15327051hci0901_2

Brown, J. S., & Duguid, P. (1998). Organizing knowledge. *California Management Review, 40*(3), 90–111. https://doi.org/10.2307/41165945

Bryman, A. (2008). *Social research methods* (3rd ed.). Oxford University Press.

Brynjolfsson, E., & McAfee, A. (2014). *The second machine age: Work, progress, and prosperity in a time of brilliant technologies.* W. W. Norton & Company.

Burke, J. L., Prewett, M. S., Gray, A. A., Yang, L., Stilson, F. R. B., Coovert, M. D., Elliot, L. R., & Redden, E. (2006). Comparing the effects of visual-auditory and visual-tactile feedback on user performance: A meta-analysis. *Proceedings of the 8th International Conference on Multimodal Interfaces,* 108–117. https://doi.org/10.1145/1180995.1181017

Burkhard, R. A., & Meier, M. (2005). Tube map visualization: Evaluation of a novel knowledge visualization application for the transfer of knowledge in long-term projects. *Journal of Universal Computer Science, 11*(4), 473–494. https://doi.org/10.3217/jucs-011-04-0473

Büttner, S., Funk, M., Sand, O., & Röcker, C. (2016). Using head-mounted displays and in-situ projection for assistive systems: A comparison. *Proceedings of the ACM International Conference on Interactive Surfaces and Spaces,* 85–94. https://doi.org/10.1145/2910674.2910679

Büttner, S., Prilla, M., & Röcker, C. (2020). Augmented reality training for industrial assembly work: Are projection-based AR assistive systems an appropriate tool for assembly training? *Proceedings of the Conference on Human Factors in Computing Systems,* 1–12. https://doi.org/10.1145/3313831.3376720

Button, G., & Harper, R. (1995). The relevance of "work-practice" for design. *Computer Supported Cooperative Work, 4*(4), 263–280. https://doi.org/10.1007/BF01846695

Button, G., & Sharrock, W. (1997). The production of order and the order of production: Possibilities for distributed organisations, work and technology in the print industry. *Proceedings of the Fifth European Conference on Computer Supported Cooperative Work,* 1–16. https://doi.org/10.1007/978-94-015-7372-6_1

Cakmakci, M. (2009). Process improvement: Performance analysis of the setup time reduction-SMED in the automobile industry. *International Journal of Advanced Manufacturing Technology, 41*(1–2), 168–179. https://doi.org/10.1007/s00170-008-1434-4

Calo, T. J. (2008). Talent management in the era of the aging workforce: The critical role of knowledge transfer. *Public Personnel Management, 37*(4), 403–416. https://doi.org/10.1177/009102600803700403

Carroll, J. M. (2000). Five reasons for scenario-based design. *Interacting with Computers, 13*(1), 43–60. https://doi.org/10.1016/S0953-5438(00)00023-0

Carstensen, P., Schmidt, K., & Wiil, U. K. (1999). Supporting shop floor intelligence: A CSCW approach to production planning and control in flexible manufacturing. In S. C. Hayne (Ed.), *Proceedings of the International ACM SIGGROUP Conference on Supporting Group Work* (pp. 111–120). ACM.

Caudell, T. P., & Mizell, D. W. (1992). Augmented reality: An application of heads-up display technology to manual manufacturing processes. *Proceedings of the Twenty-Fifth Hawaii International Conference on System Sciences, 2,* 659–669. https://doi.org/10.1109/HICSS.1992.183317

Chandler, P., & Sweller, J. (1991). Cognitive load theory and the format of instruction. *Cognition and Instruction, 8*(4), 293–332. https://doi.org/10.1207/s1532690xci0804_2

Chen, T., & Tsai, H.-R. (2017). Ubiquitous manufacturing: Current practices, challenges, and opportunities. *Robotics and Computer-Integrated Manufacturing, 45,* 126–132. https://doi.org/10.1016/j.rcim.2016.01.001

Chen, Y. C., Shang, R. A., & Kao, C. Y. (2009). The effects of information overload on consumers' subjective state towards buying decision in the internet shopping environment. *Electronic Commerce Research and Applications, 8*(1), 48–58. https://doi.org/10.1016/j.elerap.2008.09.001

Chen, Z., Li, J., Hua, Y., Shen, R., & Basu, A. (2017). Multimodal interaction in augmented reality. *2017 IEEE International Conference on Systems, Man, and Cybernetics (SMC 2017),* 206–209. https://doi.org/10.1109/SMC.2017.8122603

Chong, C., Kumar, S. P., & Member, S. (2003). Sensor networks: Evolution, opportunities, and challenges. *Proceedings of the IEEE, 91*(8), 1247–1256.

Chung, K. H., Shewchuk, J. P., & Williges, R. C. (1999). Application of augmented reality to thickness inspection. *Human Factors and Ergonomics in Manufacturing, 9*(4), 331–342. https://doi.org/10.1002/(SICI)1520-6564(199923)9:4<331::AID-HFM1>3.0.CO;2-3

Cimini, C., Pirola, F., Pinto, R., & Cavalieri, S. (2020). A human-in-the-loop manufacturing control architecture for the next generation of production systems. *Journal of Manufacturing Systems, 54,* 258–271. https://doi.org/10.1016/j.jmsy.2020.01.002

Clarke, K., Hughes, J., Martin, D., Rouncefield, M., Sommerville, I., Gurr, C., Hartswood, M., Procter, R., Slack, R., & Voss, A. (2003). Dependable red hot action. *Proceedings of the 2003 European CSCW Conference (ECSCW '03),* 14–18.

Conrad, R. W. (2016). Single Minute Exchange of Die (SMED). In Institut für angewandte Arbeitswissenschaft (Ed.), *5S als Basis des kontinuierlichen Verbesserungsprozesses* (pp. 51–56). Springer Vieweg. https://doi.org/10.1007/978-3-662-48552-1_8

Corbett, J. M. (1990). Human centred advanced manufacturing systems: From rhetoric to reality. *International Journal of Industrial Ergonomics, 5*(1), 83–90. https://doi.org/10.1016/0169-8141(90)90030-6

Crabtree, A., O'Neill, J., Tolmie, P., Castellani, S., Colombino, T., & Grasso, A. (2006). The practical indispensability of articulation work to immediate and remote help-giving. *Proceedings of the 2006 20th Anniversary Conference on Computer Supported Cooperative Work (CSCW '06),* 219–228. https://doi.org/10.1145/1180875.1180910

Crabtree, A., & Rodden, T. (2002). Ethnography and design. *International Workshop on 'Interpretive' Approaches to Information Systems and Computing Research,* 70–74.

Creswell, J. W. (2014). *Research design: Qualitative, quantitative, and mixed methods approaches* (4th ed.). SAGE Publications Inc.

Crossley, N. (1995). Merleau-Ponty, the elusive body and carnal sociology. *Body & Society, 1*(1), 43–63. https://doi.org/10.1177/1357034X95001001004

Czarniawska, B. (2007). *Shadowing and other techniques for doing fieldwork in modern societies.* Copenhagen Business School Press.

Davis, J., Edgar, T., Porter, J., Bernaden, J., & Sarli, M. (2012). Smart manufacturing, manufacturing intelligence and demand-dynamic performance. *Computers & Chemical Engineering, 47,* 145–156. https://doi.org/10.1016/J.COMPCHEMENG.2012.06.037

De Mickelis, G., & Grasso, M. A. (1994). Situating conversations within the language/action perspective: The Milan conversation model. *Proceedings of the 1994 ACM Conference on Computer Supported Cooperative Work (CSCW '94),* 89–100. https://doi.org/10.1145/192844.192880

Denef, S., & Keyson, D. (2012). Talking about implications for design in pattern language. *Proceedings of the 2012 ACM Annual Conference on Human Factors in Computing Systems (CHI '12),* 2509–2518. https://doi.org/10.1145/2207676.2208418

Dey, A., Billinghurst, M., Lindeman, R. W., & Swan, J. E. (2018). A systematic review of 10 years of augmented reality usability studies: 2005 to 2014. *Frontiers in Robotics and AI, 5*(April), 1–28. https://doi.org/10.3389/frobt.2018.00037

Diggins, T., & Tolmie, P. (2003). The "adequate" design of ethnographic outputs for practice: Some explorations of the characteristics of design resources. *Personal and Ubiquitous Computing, 7*(3–4), 147–158.

Dillon, A. P., & Shingo, S. (1985). *A revolution in manufacturing: The SMED system.* CRC Press.

Divitini, M., Sale, G. O., Pozzoli, A., & Simone, C. (1993). Supporting the dynamics of knowledge sharing within organizations. *Proceedings of the Conference on Organizational Computing Systems,* 178–183. https://doi.org/10.1145/168555.168574

Doetsch, S. (2016). *Wissenstransfer bei der Reintegration von Expatriates: Theoretische und empirische Analyse unternehmensinterner Strukturen und Prozesse.* Springer-Verlag.

Döring, H. (2016). *Wissensmanagement in Familienunternehmen.* Springer Fachmedien Wiesbaden GmbH. https://doi.org/10.1007/978-3-658-13681-9

Dourish, P. (2004). What we talk about when we talk about context. *Personal and Ubiquitous Computing, 8*(1), 19–30. https://doi.org/10.1007/s00779-003-0253-8

Dregger, J., Niehaus, J., Ittermann, P., Hirsch-Kreinsen, H., & Ten Hompel, M. (2016). The digitization of manufacturing and its societal challenges: A framework for the future of industrial labor. *2016 IEEE International Symposium on Ethics in Engineering, Science and Technology*. https://doi.org/10.1109/ETHICS.2016.7560045

Duan, J., Yu, S., Tan, H. L., Zhu, H., & Tan, C. (2022). A Survey of Embodied AI: From Simulators to Research Tasks. *IEEE Transactions on Emerging Topics in Computational Intelligence, 6*(2), 230–244. https://doi.org/10.1109/TETCI.2022.3141105

Dunston, P. S., & Shin, D. H. (2009). Key areas and issues for augmented reality applications on construction sites. In X. Wang & M. A. Schnabel (Eds.), *Mixed Reality in Architecture, Design and Construction* (pp. 157–170). Springer Netherlands. https://doi.org/10.1007/978-1-4020-9088-2_10

Echtler, F., Sturm, F., Kindermann, K., Klinker, G., Stilla, J., Trilk, J., & Najafi, H. (2004). The intelligent welding gun: Augmented reality for experimental vehicle construction. *Virtual and Augmented Reality Applications in Manufacturing*, 333–360. https://doi.org/10.1007/978-1-4471-3873-0_17

Eder, M., Hulla, M., Mast, F., & Ramsauer, C. (2020). On the application of augmented reality in a learning factory working environment. *Procedia Manufacturing, 45*, 7–12. https://doi.org/10.1016/j.promfg.2020.04.030

Ellis, C., & Wainer, J. (1994). A conceptual model of groupware. *Proceedings of the 1994 ACM Conference on Computer Supported Cooperative Work, CSCW 1994*, 79–88. https://doi.org/10.1145/192844.192878

Emmanouilidis, C., Pistofidis, P., Bertoncelj, L., Katsouros, V., Fournaris, A., Koulamas, C., & Ruiz-Carcel, C. (2019). Enabling the human in the loop: Linked data and knowledge in industrial cyber-physical systems. *Annual Reviews in Control, 47*, 249–265. https://doi.org/10.1016/j.arcontrol.2019.03.004

Engelbart, D. (1988). The augmented knowledge workshop. *A History of Personal Workstations*, 185–248.

Engelke, T., Keil, J., Rojtberg, P., Wientapper, F., Webel, S., & Bockholt, U. (2013). Content first—A concept for industrial augmented reality maintenance applications using mobile devices. *2013 IEEE International Symposium on Mixed and Augmented Reality (ISMAR), 2*, 251–252. https://doi.org/10.1109/ISMAR.2013.6671790

Evans, G., Miller, J., Iglesias Pena, M., MacAllister, A., & Winer, E. (2017). Evaluating the Microsoft HoloLens through an augmented reality assembly application. *Degraded Environments: Sensing, Processing, and Display 2017, 10197*. https://doi.org/10.1117/12.2262626

Fakourfar, O., Ta, K., Tang, R., Bateman, S., & Tang, A. (2016). Stabilized annotations for mobile remote assistance. *Proceedings of the 2016 CHI Conference on Human Factors in Computing Systems—CHI '16*, 1548–1560. https://doi.org/10.1145/2858036.2858171

Feinberg, S., & Murphy, M. (2000). Applying cognitive load theory to the design of Web-based instruction. *IPCC/SIGDOC 2000: Technology and Teamwork—Proceedings, IEEE Professional Communication Society International Professional Communication Conference and ACM Special Interest Group on Documentation Conference*, 353–360. https://doi.org/10.1109/IPCC.2000.887293

Feiner, S., Macintyre, B., & Seligmann, D. (1993). Knowledge-based augmented reality. *Communications of the ACM, 36*(7), 53–62. https://doi.org/10.1145/159544.159587

Fernández del Amo, I., Erkoyuncu, J. A., Roy, R., Palmarini, R., & Onoufriou, D. (2018). A systematic review of Augmented Reality content-related techniques for knowledge transfer in maintenance applications. *Computers in Industry, 103*, 47–71. https://doi.org/10.1016/j.compind.2018.08.007

Ferradás, P. G., & Salonitis, K. (2013). Improving changeover time: A tailored SMED approach for welding cells. *Procedia CIRP, 7*, 598–603. https://doi.org/10.1016/j.procir.2013.06.039

Fillatreau, P., Fourquet, J. Y., Le Bolloc'H, R., Cailhol, S., Datas, A., & Puel, B. (2013). Using virtual reality and 3D industrial numerical models for immersive interactive checklists. *Computers in Industry, 64*(9), 1253–1262. https://doi.org/10.1016/j.compind.2013.03.018

Finkenzeller, K., & Waddington, R. (2003). *RFID Handbook: Fundamentals and Applications in Contactless Smart Cards and Identification*. Wiley. https://doi.org/10.1002/9780470665121

Fiorentino, M., Uva, A. E., Gattullo, M., Debernardis, S., & Monno, G. (2014). Augmented reality on large screen for interactive maintenance instructions. *Computers in Industry, 65*(2), 270–278. https://doi.org/10.1016/j.compind.2013.11.004

Fitzpatrick, G. (2003). Emergent expertise sharing in a new community. In M. S. Ackerman, V. Pipek, & V. Wulf (Eds.), *Sharing Expertise: Beyond Knowledge Management* (pp. 81–110). The MIT Press.

Fu, W. T. (2008). The microstructures of social tagging: A rational model. *Proceedings of the ACM Conference on Computer Supported Cooperative Work, CSCW*, 229–238. https://doi.org/10.1145/1460563.1460600

Funk, M., Bächler, A., Bächler, L., Kosch, T., Heidenreich, T., & Schmidt, A. (2017). Working with Augmented Reality? *Proceedings of the ACM International Joint Conference on Pervasive and Ubiquitous Computing, UbiComp*, 222–229. https://doi.org/10.1145/3056540.3056548

Funk, M., Kosch, T., Greenwald, S. W., & Schmidt, A. (2015). A benchmark for interactive augmented reality instructions for assembly tasks. *Proceedings of the 14th International Conference on Mobile and Ubiquitous Multimedia*, 253–257. https://doi.org/10.1145/2836041.2836067

Funk, M., Kosch, T., & Schmidt, A. (2016). Interactive worker assistance: Comparing the effects of in-situ projection, head-mounted displays, tablet, and paper instructions. *Proceedings of the 2016 ACM International Joint Conference on Pervasive and Ubiquitous Computing, UbiComp*, 934–939. https://doi.org/10.1145/2971648.2971706

Fussell, S. R., Setlock, L. D., & Kraut, R. E. (2003). Effects of head-mounted and scene-oriented video systems on remote collaboration on physical tasks. *Proceedings of the Conference on Human Factors in Computing Systems—CHI '03*, 513. https://doi.org/10.1145/642611.642701

Gallagher, S. (2010). Merleau-Ponty's Phenomenology of Perception. *Topoi, 29*(2), 183–185. https://doi.org/10.1007/s11245-010-9079-y

Gattullo, M., Scurati, G. W., Fiorentino, M., Uva, A. E., Ferrise, F., & Bordegoni, M. (2019). Towards augmented reality manuals for industry 4.0: A methodology. *Robotics and Computer-Integrated Manufacturing, 56*(October 2018), 276–286. https://doi.org/10.1016/j.rcim.2018.10.001

Gauglitz, S., Lee, C., Turk, M., & Höllerer, T. (2012). Integrating the physical environment into mobile remote collaboration. *Proceedings of the 14th International Conference on Human-Computer Interaction with Mobile Devices and Services*, 241–250.

Gauglitz, S., Nuernberger, B., Turk, M., & Höllerer, T. (2014a). In touch with the remote world. *Proceedings of the 20th ACM Symposium on Virtual Reality Software and Technology—VRST '14*, 197–205. https://doi.org/10.1145/2671015.2671016

Gauglitz, S., Nuernberger, B., Turk, M., & Höllerer, T. (2014b). World-stabilized annotations and virtual scene navigation for remote collaboration. *Proceedings of the 27th Annual ACM Symposium on User Interface Software and Technology—UIST '14*, 449–459. https://doi.org/10.1145/2642918.2647372

Gellersen, H. (2005). Smart-Its: Computers for artifacts in the physical world. *Communications of the ACM, 48*(3), 66–66. https://doi.org/10.1145/1047671.1047707

Gibson, W., & Brown, A. (2009). *Working with qualitative data.* SAGE. https://doi.org/10.1007/s13398-014-0173-7.2

Goh, S. C. (2002). Managing effective knowledge transfer: An integrative framework and some practice implications. *Journal of Knowledge Management, 6*(12), 18–35.

Gonzalez, A. V., Koh, S., Kapalo, K., Sottilare, R., Garrity, P., Billinghurst, M., & Laviola, J. (2019). A comparison of desktop and augmented reality scenario-based training authoring tools. *Proceedings—2019 IEEE International Symposium on Mixed and Augmented Reality, ISMAR 2019*, 339–350. https://doi.org/10.1109/ISMAR.2019.00032

Goodwin, C. (1994). Professional vision. *American Anthropologist.* https://doi.org/10.1525/aa.1994.96.3.02a00100

Goomas, D. T. (2012). Immediate feedback on accuracy and performance: The effects of wireless technology on food safety tracking at a distribution center. *Journal of Organizational Behavior Management, 32*(4), 320–328. https://doi.org/10.1080/01608061.2012.729391

Goomas, D. T., Smith, S. M., & Ludwig, T. D. (2011). Business activity monitoring: Real-time group goals and feedback using an overhead scoreboard in a distribution center. *Journal of Organizational Behavior Management, 31*(3), 196–209. https://doi.org/10.1080/01608061.2011.589715

Gorecky, D., Schmitt, M., Loskyll, M., & Zühlke, D. (2014). Human-machine-interaction in the industry 4.0 era. *Proceedings—2014 12th IEEE International Conference on Industrial Informatics, INDIN 2014*, 289–294. https://doi.org/10.1109/INDIN.2014.6945523

Goubergen, D. Van, & Lockhart, T.E. (2005). Human Factors Aspects in Set-Up Time Reduction. In: Zülch, G., Jagdev, H.S., Stock, P. (eds) *Integrating Human Aspects in Production Management.* IFIP International Conference for Information Processing, vol 160. Springer, Boston, MA. https://doi.org/10.1007/0-387-23078-5_10

Goubergen, D. Van, & Landeghem, H. Van. (2002). Rules for integrating fast changeover capabilities into new equipment design. Robotics and Computer-Integrated Manufacturing, 18(3–4), 205–214. https://doi.org/10.1016/S0736-5845(02)00011-X

Grandi, F., Peruzzini, M., Zanni, L., Campanella, C. E., & Pellicciari, M. (2018). Digital manufacturing and virtual reality for tractors' human-centred design. *Advances in*

Transdisciplinary Engineering, 7, 702–711. https://doi.org/10.3233/978-1-61499-898-3-702
Grandi, F., Zanni, L., Peruzzini, M., Pellicciari, M., & Campanella, C. E. (2020). A transdisciplinary digital approach for tractor's human-centred design. *International Journal of Computer Integrated Manufacturing, 33*(4), 377–397. https://doi.org/10.1080/0951192X.2019.1599441
Grant, R. M. (1996). Toward a knowledge-based theory of the firm. *Strategic Management Journal, 17*(S2), 109–122. https://doi.org/10.1002/smj.4250171110
Grant, R. M., & Baden-Fuller, C. (2004). A knowledge accessing theory of strategic alliances. *Journal of Management Studies, 41*(1), 61–84. https://doi.org/10.1111/j.1467-6486.2004.00421.x
Greatbatch, D., Luff, P., Heath, C., & Campion, P. (1993). Interpersonal communication and human-computer interaction: An examination of the use of computers in medical consultations. *Interacting with Computers*. https://doi.org/10.1016/0953-5438(93)90018-O
Guba, E. G. (1981). Criteria for assessing the trustworthiness of naturalistic inquiries. *ECTJ, 29*(2), 75–91. https://doi.org/10.1007/BF02766777
Guerreiro, B. V., Lins, R. G., Sun, J., & Schmitt, R. (2018). Definition of smart retrofitting: First steps for a company to deploy aspects of Industry 4.0. In A. Hamrol, O. Ciszak, S. Legutko, & M. Jurczyk (Eds.), *Advances in Manufacturing* (pp. 161–170). Springer International Publishing. https://doi.org/10.1007/978-3-319-68619-6_16
Gupta, B., Iyer, L., & Aronson, J. (2000). Knowledge management: Practices and challenges. *Industrial Management & Data Systems, 100*(1), 17–21. http://www.emeraldinsight.com/doi/abs/https://doi.org/10.1108/02635570010273018
Hager, P. (2000). Know-how and workplace practical judgement. *Journal of the Philosophy of Education, 34*(2), 281–296. https://doi.org/10.1111/1467-9752.00173
Haller, M. (2004). Photorealism or/and non-photorealism in augmented reality. *Proceedings of the 2004 ACM SIGGRAPH International Conference on Virtual Reality Continuum and Its Applications in Industry—VRCAI '04*, 189. https://doi.org/10.1145/1044588.1044627
Harper, R., Procter, R., Randall, D., & Rouncefield, M. (2001). "Safety in numbers" calculation and document re-use in knowledge work. *Proceedings of the 2001 ACM International Conference on Supporting Group Work*. https://doi.org/10.1145/500286.500322
Harper, R., Randall, D., & Rouncefield, M. (2000). *Organisational change and retail finance: An ethnographic perspective*. Routledge.
Harper, R., Rodden, T., Rogers, Y., & Sellen, A. (2008). *Being human: Human-computer interaction in the year 2020*. Microsoft Research Ltda.
Harrison, S., & Minneman, S. (1995). Studying collaborative design to build design tools. *CAAD Futures Digital Proceedings 1995*, 687–697.
Hart, S. G., & Staveland, L. E. (1988). Development of NASA-TLX (Task Load Index): Results of empirical and theoretical research. In *Human Mental Workload. Advances in Psychology* (Issue 52, pp. 139–183). https://doi.org/10.1016/S0166-4115(08)62386-9
Hau, Y. S., Kim, B., Lee, H., & Kim, Y. G. (2013). The effects of individual motivations and social capital on employees' tacit and explicit knowledge sharing intentions. *International Journal of Information Management, 33*(2), 356–366. https://doi.org/10.1016/j.ijinfomgt.2012.10.009

Heath, C., & Luff, P. (1991). Disembodied conduct: Communication through video in a multi-media office environment. *Proceedings of the SIGCHI Conference on Human Factors in Computing Systems*. https://doi.org/10.1145/108844.108859

Heath, C., Luff, P., & Sanchez Svensson, M. (2007). Video and qualitative research: Analysing medical practice and interaction. *Medical Education, 41*(1), 109–116. https://doi.org/10.1111/j.1365-2929.2006.02641.x

Heath, C., Luff, P., & Sellen, A. (1995). Reconsidering the virtual workplace: Flexible support for collaborative activity. *Proceedings of the Fourth European Conference on Computer-Supported Cooperative Work ECSCW '95*, 83–99.

Heidemann, G., Bax, I., & Bekel, H. (2004). Multimodal interaction in an augmented reality scenario. *Proceedings of the 6th International Conference on Multimodal Interfaces—ICMI '04*, 2017-Janua, 53. https://doi.org/10.1145/1027933.1027944

Henderson, S., & Feiner, S. (2011). Exploring the benefits of augmented reality documentation for maintenance and repair. *IEEE Transactions on Visualization and Computer Graphics, 17*(10), 1355–1368. https://doi.org/10.1109/TVCG.2010.245

Hendriks, P. (1999). Why share knowledge? The influence of ICT on the motivation for knowledge sharing. *Knowledge and Process Management, 6*(2), 91–100. https://doi.org/10.1002/(SICI)1099-1441(199906)6:2<91::AID-KPM54>3.0.CO;2-M

Henrysson, A., Ollila, M., & Billinghurst, M. (2005). Mobile phone based AR scene assembly. *Proceedings of the 4th International Conference on Mobile and Ubiquitous Multimedia—MUM '05*, 95. https://doi.org/10.1145/1149488.1149504

Hermanowicz, J. C. (2002). The great interview: 25 strategies for studying people in bed. *Qualitative Sociology, 25*(4), 479–499. https://doi.org/10.1023/A:1021062932081

Hermsen, S., Frost, J., Renes, R. J., & Kerkhof, P. (2016). Using feedback through digital technology to disrupt and change habitual behavior: A critical review of current literature. *Computers in Human Behavior, 57*, 61–74. https://doi.org/10.1016/j.chb.2015.12.023

Herterich, M. M., Uebernickel, F., & Brenner, W. (2015). The impact of cyber-physical systems on industrial services in manufacturing. *Procedia CIRP, 30*, 323–328. https://doi.org/10.1016/j.procir.2015.02.110

Hickley, A. (2003). Verbatim, intelligent verbatim or edited transcription? https://www.penguin-transcription.co.uk/transcription-type-verbatim-intelligent-verbatim-or-edited/

Hiltz, S. R., & Turoff, M. (1985). Structuring computer-mediated communication systems to avoid information overload. *Communications of the ACM, 28*(7), 680–689. https://doi.org/10.1145/3894.3895

Hindmarsh, J., & Pilnick, A. (2002). The tacit order of teamwork: Collaboration and embodied conduct in anesthesia. *Sociological Quarterly*. https://doi.org/10.1111/j.1533-8525.2002.tb00044.x

Hirsch-Kreinsen, H. (2016). Digitization of industrial work: Development paths and prospects. *Journal for Labour Market Research, 49*(1), 1–14. https://doi.org/10.1007/s12651-016-0200-6

Hoffmann, A. (2009). Entwicklung eines Ordnungsrahmens zur Analyse von intraorganisationalem Wissenstransfer. In *Berichte des Fachbereichs Wirtschaftswissenschaften der Goethe-Universität Frankfurt am Main.*

Holtzblatt, K., & Jones, S. (2019). Contextual inquiry: A participatory technique for system design. In *Participatory Design* (pp. 177–210).

Howells, J. R. L. (1996a). Tacit knowledge, innovation and technology transfer. *Technology Analysis and Strategic Management, 8*(2), 91–106.

Hsieh, G., Hudson, S. E., & Kraut, R. E. (2011). Donate for credibility: How contribution incentives can improve credibility. *Group*, 3435–3438. https://doi.org/10.1145/1978942.1979454

Hu, F., Lu, Y., Vasilakos, A. V., Hao, Q., Ma, R., Patil, Y., Zhang, T., Lu, J., Li, X., & Xiong, N. N. (2016). Robust cyber-physical systems: Concept, models, and implementation. *Future Generation Computer Systems, 56*, 449–475. https://doi.org/10.1016/j.future.2015.06.006

Huang, J., Mei, H., Tang, Z., Li, J., Zhang, X., Lu, Y., Huang, F., Jin, Q., & Wang, Z. (2017). Triple-amiRNA VEGFRs inhibition in pancreatic cancer improves the efficacy of chemotherapy through EMT regulation. *Journal of Controlled Release, 245*, 1–14. https://doi.org/10.1016/j.jconrel.2016.11.024

Irlitti, A., Smith, R. T., Von Itzstein, S., Billinghurst, M., & Thomas, B. H. (2017). Challenges for asynchronous collaboration in augmented reality. *Adjunct Proceedings of the 2016 IEEE International Symposium on Mixed and Augmented Reality, ISMAR-Adjunct 2016*, 31–35. https://doi.org/10.1109/ISMAR-Adjunct.2016.0032

Janlert, L.-E., & Stolterman, E. (2010). Complex interaction. *ACM Transactions on Computer-Human Interaction, 17*(2), 1–32. https://doi.org/10.1145/1746259.1746262

Jarrahi, M. H., Askay, D., Eshraghi, A., & Smith, P. (2023). Artificial intelligence and knowledge management: A partnership between human and AI. *Business Horizons, 66*(1), 87–99. https://doi.org/10.1016/J.BUSHOR.2022.03.002

Jasche, F., & Ludwig, T. (2020). PrintARface: Supporting the exploration of cyber-physical systems through augmented reality. *Proceedings of the 11th Nordic Conference on Human-Computer Interaction: Shaping Experiences, Shaping Society*, 1–12. https://doi.org/10.1145/3419249.3420162

Jeschke, S., Brecher, C., Meisen, T., Özdemir, D., & Eschert, T. (2017). Industrial internet of things and cyber manufacturing systems. 3–19. https://doi.org/10.1007/978-3-319-42559-7_1

Jetter, J., Eimecke, J., & Rese, A. (2018). Augmented reality tools for industrial applications: What are potential key performance indicators and who benefits? *Computers in Human Behavior, 87*, 18–33. https://doi.org/10.1016/j.chb.2018.04.054

Jo, G. S., Oh, K. J., Ha, I., Lee, K. S., Hong, M. D., Neumann, U., & You, S. (2014). A unified framework for augmented reality and knowledge-based systems in maintaining aircraft. *Proceedings of the National Conference on Artificial Intelligence, 4*, 2990–2997.

Jönsson, A., Wall, J., & Broman, G. (2005). A virtual machine concept for real-time simulation of machine tool dynamics. *International Journal of Machine Tools and Manufacture, 45*(7–8), 795–801. https://doi.org/10.1016/J.IJMACHTOOLS.2004.11.012

Kagermann, H., Wahlster, W., & Helbig, J. (2013). Recommendations for implementing the strategic initiative INDUSTRIE 4.0. *Final Report of the Industrie 4.0 WG, April, 82*. https://doi.org/10.13140/RG.2.1.1205.8966

Kalyuga, S. (2008). *Managing cognitive load in adaptive multimedia learning.*

Kalyuga, S., Ayres, P., Chandler, P., & Sweller, J. (2003). The expertise reversal effect. *Educational Psychologist, 38*, 23–31. https://doi.org/10.1207/S15326985EP3801_4

Keller, T., & Tergan, S.-O. (2005). Visualizing knowledge and information: An introduction. In S.-O. Tergan & T. Keller (Eds.), *Knowledge and information visualization: Searching synergies* (pp. 1–23). Springer-Verlag.

Kendon, A. (1990). Spatial organization in social encounters. In Kendon, A. *Conducting interaction: Patterns of behavior in focused encounter*.

Klinker, G., Creighton, O., Dutoit, A. H., Kobylinski, R., Vilsmeier, C., & Brügge, B. (2001). Augmented maintenance of power plants: A prototyping case study of a mobile AR system. *Proc. ISAR 2001*, 124–133. https://doi.org/10.1109/ISAR.2001.970522

Klinker, G., Najafi, H., Sielhorst, T., Sturm, F., Echtler, F., Isik, M., Wein, W., & Trübswetter, C. (2004). FixIt: An approach towards assisting workers in diagnosing machine malfunctions. *Proc. of the International Workshop Exploring the Design and Engineering of Mixed Reality Systems—MIXER 2004.*

Klopfer, E., Perry, J., Squire, K., & Jan, M. (2005). Collaborative learning through augmented reality role playing. *Collaborative Learning through Augmented Reality Role Playing*, 311–315. https://doi.org/10.3115/1149293.1149333

Knoblauch, H., Baer, A., Laurier, E., Petschke, S., & Schnettler, B. (2008). Visual analysis: New developments in the interpretative analysis of video and photography. *Forum Qualitative Sozialforschung / Forum: Qualitative Social Research, 9*(3). http://www.qualitative-research.net/index.php/fqs/article/view/1170

Korn, O., Schmidt, A., & Hörz, T. (2013). The potentials of in-situ-projection for augmented workplaces in production: A study with impaired persons. *Conference on Human Factors in Computing Systems—Proceedings, 2013-April*, 979–984. https://doi.org/10.1145/2468356.2468531

Koschmann, T., LeBaron, C., Goodwin, C., Zemel, A., & Dunnington, G. (2007). Formulating the triangle of doom. *Gesture*. https://doi.org/10.1075/gest.7.1.06kos

Kosslyn, S. M. (2006). *Graph design for the eye and mind*. Oxford University Press.

Kumar, K., Divya, & Davim, Z. J. P. (2018). *Advanced machining and manufacturing processes*. Springer Nature. https://doi.org/10.1007/978-3-319-76075-9

Langlotz, T., Mooslechner, S., Zollmann, S., Degendorfer, C., Reitmayr, G., & Schmalstieg, D. (2010). Sketching up the world: In-situ authoring for mobile augmented reality. *Personal and Ubiquitous Computing*, 1–8. https://doi.org/10.1007/s00779-011-0430-0

Lapre, M. A., Mukherjee, A. S., & Van Wassenhove, L. N. (2000). Behind the learning curve: Linking learning activities to waste reduction. *Management Science, 46*(5), 597–611. https://doi.org/10.1287/mnsc.46.5.597.12049

Laugwitz, B., Held, T., & Schrepp, M. (2008). Construction and evaluation of a user experience questionnaire. *Lecture Notes in Computer Science*, 63–76. https://doi.org/10.1007/978-3-540-89350-9_6

Lave, J., & Wenger, E. (1991). *Situated learning: Legitimate peripheral participation*. Cambridge University Press. https://doi.org/10.2307/2804509

Lee, C. P., & Paine, D. (2015). From the matrix to a model of coordinated action (MoCA): A conceptual framework of and for CSCW. *CSCW 2015—Proceedings of the 2015 ACM International Conference on Computer-Supported Cooperative Work and Social Computing*, 179–194. https://doi.org/10.1145/2675133.2675161

Lee, E. A. (2008). Cyber physical systems: Design challenges. *11th IEEE International Symposium on Object Oriented Real-Time Distributed Computing*, 10. https://doi.org/10.1109/ISORC.2008.25

Lee, E. A., & Seshia, S. A. (2015). *Introduction to embedded systems: A cyber-physical systems approach* (2nd ed.). LeeSeshia.org. https://doi.org/10.1016/B978-0-12-386886-2.00001-1

Lee, G. A., Nelles, C., Billinghurst, M., & Kim, G. J. (2004). Immersive authoring of tangible augmented reality applications. *ISMAR 2004: Proceedings of the Third IEEE and ACM International Symposium on Mixed and Augmented Reality*, 172–181. https://doi.org/10.1109/ISMAR.2004.34

Lee, J., Bagheri, B., & Kao, H. (2014). Recent advances and trends of cyber-physical systems and big data analytics in industrial informatics. *International Proceeding of Int Conference on Industrial Informatics (INDIN)*, July. https://doi.org/10.13140/2.1.1464.1920

Lee, J., Bagheri, B., & Kao, H. A. (2015). A cyber-physical systems architecture for Industry 4.0-based manufacturing systems. *Manufacturing Letters, 3*, 18–23. https://doi.org/10.1016/j.mfglet.2014.12.001

Lee, J., Lapira, E., Bagheri, B., & Kao, H. (2013). Recent advances and trends in predictive manufacturing systems in big data environment. *Manufacturing Letters, 1*(1), 38–41. https://doi.org/10.1016/j.mfglet.2013.09.005

Lee, S., & Akin, Ö. (2011). Augmented reality-based computational fieldwork support for equipment operations and maintenance. *Automation in Construction, 20*(4), 338–352. https://doi.org/10.1016/j.autcon.2010.11.004

Leit, P., Mal, I., Sedl, D., Maly, I., Sedlácek, D., & Leitão, P. (2016). Augmented reality experiments with industrial robot in Industry 4.0 environment. *Industrial Informatics (INDIN), 14*(2378–363X), 176–181. https://doi.org/10.1109/INDIN.2016.7819154

Levin, D. Z. (2000). Organizational learning and the transfer of knowledge: An investigation of quality improvement. *Organization Science, 11*(6), 630–647. https://doi.org/10.1287/orsc.11.6.630.12535

Lewkowicz, M., & Liron, R. (2019). The missing "turn to practice" in the digital transformation of industry. *Computer Supported Cooperative Work: CSCW: An International Journal, 28*(3–4), 655–683. https://doi.org/10.1007/s10606-019-09347-y

Li, D., Landström, A., Fast-Berglund, Å., & Almström, P. (2019). Human-centred dissemination of data, information and knowledge in Industry 4.0. *Procedia CIRP, 84*, 380–386. https://doi.org/10.1016/j.procir.2019.04.261

Lisiankova, K., & Wright, R. E. (2005). Demographic change and the European Union labour market. *National Institute Economic Review, 194*(1), 74–81. https://doi.org/10.1177/0027950105061497

Liu, X. F., Shahriar, M. R., Al Sunny, S. M. N., Leu, M. C., & Hu, L. (2017). Cyber-physical manufacturing cloud: Architecture, virtualization, communication, and testbed. *Journal of Manufacturing Systems, 43*, 352–364. https://doi.org/10.1016/j.jmsy.2017.04.004

Ludwig, T., Boden, A., & Pipek, V. (2017). 3D printers as sociable technologies: Taking appropriation infrastructures to the Internet of Things. *ACM Transactions on Computer-Human Interaction, 24*(2). https://doi.org/10.1145/3007205

Ludwig, T., Döll, M., & Kotthaus, C. (2019). "The printer is telling me about itself": Supporting the appropriation of hardware by using projection mapping. *DIS 2019—Proceedings of the 2019 ACM Designing Interactive Systems Conference*, 331–344. https://doi.org/10.1145/3322276.3322342

Ludwig, T., Kotthaus, C., & Pipek, V. (2016). Should I try turning it off and on again? *International Journal of Information Systems for Crisis Response and Management, 7*(3), 55–68. https://doi.org/10.4018/ijiscram.2015070104

Ludwig, T., Kotthaus, C., Stein, M., Durt, H., Kurz, C., Wenz, J., Doublet, T., Becker, M., Pipek, V., & Wulf, V. (2016). Arbeiten im Mittelstand 4.0—KMU im Spannungsfeld des digitalen Wandels. *HMD Praxis der Wirtschaftsinformatik, 53*(1), 1–16. https://doi.org/10.1365/s40702-015-0200-y

Ludwig, T., Stickel, O., Boden, A., & Pipek, V. (2014). Towards sociable technologies: An empirical study on designing appropriation infrastructures for 3D printing. *Designing Interactive Systems, 1*(1), 835–844. https://doi.org/10.1145/2598510.2598528

Ludwig, T., Stickel, O., Tolmie, P., & Sellmer, M. (2021). shARe-IT: Ad hoc remote troubleshooting through augmented reality. *Computer Supported Cooperative Work (CSCW), 30*(1), 119–167. https://doi.org/10.1007/s10606-021-09393-5

Ludwig, T., Tolmie, P., & Pipek, V. (2017). From the internet of things to an internet of practices. *Proceedings of 15th European Conference on Computer-Supported Cooperative Work—Exploratory Papers*. https://doi.org/10.18420/ecscw2017-10

Ludwig, T., Kotthaus, C., Stein, M., Pipek, V., & Wulf, V. (2018). Revive old discussions! Socio-technical challenges for small and medium enterprises within Industry 4.0. *Proceedings of the 16th European Conference on Computer-Supported Cooperative Work—Exploratory Papers*, Reports of the European Society for Socially Embedded Technologies. https://doi.org/10.18420/ecscw2018_15

Luff, P., Hindmarsh, J., & Heath, C. (2000). Workplace studies. In P. Luff, J. Hindmarsh, & C. Heath (Eds.), *Workplace studies: Recovering work practice and informing system design* (pp. 1–25). Cambridge University Press. https://doi.org/10.1017/CBO9780511628122

Lynch, M. (1997). *Scientific practice and ordinary action: Ethnomethodology and social studies of science*. Cambridge University Press.

MacIntyre, B., Gandy, M., Dow, S., & Bolter, J. D. (2005). DART: A toolkit for rapid design exploration of augmented reality experiences. *ACM Transactions on Graphics, 24*(3), 932. https://doi.org/10.1145/1073204.1073288

MacKelprang, A. W., & Nair, A. (2010). Relationship between just-in-time manufacturing practices and performance: A meta-analytic investigation. *Journal of Operations Management, 28*(4), 283–302. https://doi.org/10.1016/j.jom.2009.10.002

Mainwaring, S. D., & Lee, C. P. (2017). Turnover and the model of coordinated action (MoCA). *CSCW 2017—Companion of the 2017 ACM Conference on Computer Supported Cooperative Work and Social Computing*, 239–242. https://doi.org/10.1145/3022198.3026359

Majchrzak, A., Faraj, S., Kane, G. C., & Azad, B. (2013). The contradictory influence of social media affordances on online communal knowledge sharing. *Journal of Computer-Mediated Communication, 19*(1), 38–55. https://doi.org/10.1111/jcc4.12030

Marcus, A. (2015). Design, user experience, and usability: Design discourse: 4th international conference, DUXU 2015 held as part of HCI international 2015 Los Angeles, CA, USA, August 2–7, 2015 proceedings, part I. In A. Marcus (Ed.), *Lecture notes in computer science (including subseries Lecture Notes in Artificial Intelligence and Lecture Notes in Bioinformatics)* (Vol. 9186, pp. 75–85). Springer International Publishing. https://doi.org/10.1007/978-3-319-20886-2

Marseu, E., Kolberg, D., Birtel, M., & Zühlke, D. (2016). Interdisciplinary engineering methodology for changeable cyber-physical production systems. *IFAC-PapersOnLine, 49*(31), 85–90. https://doi.org/10.1016/j.ifacol.2016.12.166

Martin, D., Rouncefield, M., & Sommerville, I. (2002). Applying patterns of cooperative interaction to work (re)design. *Proceedings of the SIGCHI Conference on Human Factors in Computing Systems*, 235–242. https://doi.org/10.1145/503376.503419

Masoni, R., Ferrise, F., Bordegoni, M., Gattullo, M., Uva, A. E., Fiorentino, M., Carrabba, E., & Di Donato, M. (2017). Supporting remote maintenance in Industry 4.0 through augmented reality. *Procedia Manufacturing, 11*(November 2018), 1296–1302. https://doi.org/10.1016/j.promfg.2017.07.257

Matthew, C. T., & Sternberg, R. J. (2009). Developing experience-based (tacit) knowledge through reflection. *Learning and Individual Differences, 19*(4), 530–540. https://doi.org/10.1016/j.lindif.2009.07.001

McIntosh, R. I., Culley, S. J., Mileham, A. R., & Owen, G. W. (2001). Changeover improvement: A maintenance perspective. *International Journal of Production Economics, 73*(2), 153–163. https://doi.org/10.1016/S0925-5273(00)00170-5

Mengis, J., & Eppler, M. J. (2012). Visualizing instead of overloading: Exploring the promise and problems of visual communication to reduce information overload. In *Information overload* (pp. 203–229). John Wiley & Sons, Inc. https://doi.org/10.1002/9781118360491.ch10

Mengoni, M., Ceccacci, S., Generosi, A., & Leopardi, A. (2018). Spatial augmented reality: An application for human work in smart manufacturing environment. *Procedia Manufacturing, 17*, 476–483. https://doi.org/10.1016/j.promfg.2018.10.072

Microsoft. (2020). *Dynamics 365 Guides*. https://dynamics.microsoft.com/en-us/mixed-reality/guides/

Miller, D. P., & Swain, A. D. (1987). Human error and human reliability. in Salvendy, G. (Eds): *Handbook of Human Factors and Ergonomics*, Wiley, New York, NY.

Mohr, P., Kerbl, B., Donoser, M., Schmalstieg, D., & Kalkofen, D. (2015). Retargeting technical documentation to augmented reality. *Conference on Human Factors in Computing Systems—Proceedings, 2015-April*, 3337–3346. https://doi.org/10.1145/2702123.2702490

Mondada, L. (2006). Video recording as the reflexive preservation and configuration of phenomenal features for analysis. In H. Knoblauch, B. Schnettler, J. Raab, & H.-G. Soeffner (Eds.), *Video analysis. Methodology and methods. Qualitative audiovisual data analysis in sociology* (pp. 51–67).

Mondada, L. (2012). The establishment of a common focus of attention: Issues for participants, camera(wo)men and analysts. 43–56.

Monk, A., Peter, W., Haber, J., & Davenport, L. (1993). Cooperative evaluation: A run-time guide. In *Improving your human-computer interface: A practical technique*. Prentice-Hall.

Monostori, L. (2014). Cyber-physical production systems: Roots, expectations and R&D challenges. *Procedia CIRP, 17*, 9–13. https://doi.org/10.1016/j.procir.2014.03.115

Monostori, L., Kádár, B., Bauernhansl, T., Kondoh, S., Kumara, S., Reinhart, G., Sauer, O., Schuh, G., Sihn, W., & Ueda, K. (2016). Cyber-physical systems in manufacturing. *CIRP Annals—Manufacturing Technology, 65*(2), 621–641. https://doi.org/10.1016/j.cirp.2016.06.005

Moore, R. J., Whalen, J., & Gathman, E. C. H. (2010). The work of the work order: Document practice in face-to-face service encounters. In *Organisation, interaction and practice* (pp. 172–197). Cambridge University Press.

Morris, M. R., Teevan, J., & Panovich, K. (2010). What do people ask their social networks, and why? *Proceedings of the 28th International Conference on Human Factors in Computing Systems—CHI '10*, 1739. https://doi.org/10.1145/1753326.1753587

Mourtzis, D., Angelopoulos, J., & Panopoulos, N. (2020). Recycling and retrofitting for industrial equipment based on augmented reality. *Procedia CIRP, 90*, 606–610. https://doi.org/10.1016/j.procir.2020.02.134

Müller, J. M., Buliga, O., & Voigt, K. I. (2018). Fortune favors the prepared: How SMEs approach business model innovations in Industry 4.0. *Technological Forecasting and Social Change, 132*, 2–17. https://doi.org/10.1016/J.TECHFORE.2017.12.019

Muller, M. J., & Kuhn, S. (1993). Participatory design. *Communications of the ACM, 36*(6), 24–28. https://doi.org/10.1145/153571.255960

Murphy, K. M. (2004). Imagination as joint activity: The case of architectural interaction. *Mind, Culture, and Activity*. https://doi.org/10.1207/s15327884mca1104_3

Nahapiet, J., & Goshal, S. (1998). Social capital, intellectual capital, and the organizational advantage. *Academy of Management Review, 23*(2), 242–266. https://doi.org/10.2307/259373

Nakanishi, M., Taguchi, K. I., & Okada, Y. (2010). Suggestions on the applicability of visual instructions with see-through head-mounted displays depending on the task. *Applied Ergonomics, 42*(1), 146–155. https://doi.org/10.1016/j.apergo.2010.06.002

Nardi, B. A., & Engeström, Y. (1999). A web on the wind: The structure of invisible work. *Computer Supported Cooperative Work, 8*, 1–8. https://doi.org/10.1023/A:1008694621289

National Institute of Standards and Technology, N. (2013). Foundations for innovation in cyber-physical systems. *Workshop Report*. National Institute of Standards and Technology. https://doi.org/10.1007/s13398-014-0173-7.2

Nebeling, M., & Madier, K. (2019). 360proto. *Proceedings of the 2019 CHI Conference on Human Factors in Computing Systems—CHI '19*, 1–13. https://doi.org/10.1145/3290605.3300826

Nebeling, M., Nebeling, J., Yu, A., & Rumble, R. (2018). ProtoAR. *Proceedings of the 2018 CHI Conference on Human Factors in Computing Systems—CHI '18*, 1–12. https://doi.org/10.1145/3173574.3173927

Nebeling, M., & Speicher, M. (2018). The trouble with augmented reality/virtual reality authoring tools. *2018 IEEE International Symposium on Mixed and Augmented Reality Adjunct (ISMAR-Adjunct)*, 333–337. https://doi.org/10.1109/ISMAR-Adjunct.2018.00098

Neumann, U., & Majoros, A. (1998). Cognitive, performance, and systems issues for augmented reality applications in manufacturing and maintenance. *Proceedings. IEEE 1998 Virtual Reality Annual International Symposium*, 4–11. https://doi.org/10.1109/vrais.1998.658416

Newell, F. N., Ernst, M. O., Tjan, B. S., & Bülthoff, H. H. (2014). Viewpoint dependence in visual and haptic object recognition. *Psychological Science, 12*(1), 37–42. https://doi.org/10.1111/1467-9280.00307

Nielsen, J. (1993). *Usability engineering*. Morgan-Kaufmann.

Nielsen, J., & Molich, R. (1990). Heuristic evaluation of user interfaces. In *Proceedings of the 1990 ACM SIGCHI Conference on Human Factors in Computing Systems* (pp. 249–256). ACM. https://doi.org/10.1145/97243.97281

Nilsson, S., & Johansson, B. (2008). Acceptance of augmented reality instructions in a real work setting. *Proceedings of ACM CHI 2008 Conference on Human Factors in Computing Systems, 2*, 2025–2032. https://doi.org/10.1145/1358628.1358633

Nonaka, I. (1988). Toward middle-up-down management: Accelerating information creation. *Sloan Management Review, 29*(3), 9–18.

Nonaka, I. (1991). The knowledge-creating company. *Harvard Business Review (Kindle Edi, Issue November-December 1991)*. Harvard Business Review Classics.

Nonaka, I. (1994). A dynamic theory of organizational knowledge creation. *Organization Science: A Journal of the Institute of Management Sciences, 5*(1), 14.

Nonaka, I., & Takeuchi, H. (1995). *The knowledge-creating company: How Japanese companies create the dynamics of innovation*. Oxford University Press. https://doi.org/10.1057/jibs.1996.13

Nonaka, I., & Toyama, R. (2003). The knowledge-creating theory revisited: Knowledge creation as a synthesizing process. *Knowledge Management Research & Practice, 1*, 2–10. https://doi.org/10.1057/palgrave.kmrp.8500001

Nonaka, I., Toyama, R., & Konno, N. (2000). SECI, Ba and leadership: A unified model of dynamic knowledge creation. *Long Range Planning, 33*(1), 5–34. https://doi.org/10.1016/S0024-6301(99)00115-6

Nonaka, I., & Von Krogh, G. (2009). Tacit knowledge and knowledge conversion: Controversy and advancement in organizational knowledge creation theory. *Organization Science, 20*(3), 635–652. https://doi.org/10.1287/orsc.1080.0412

Norman, D. A. (1998). *The design of everyday things* (2nd ed.). Doubleday. https://doi.org/10.1002/hfm.20127

Normark, M., & Randall, D. (2005). Local expertise at an emergency call centre. *ECSCW 2005: Proceedings of the Ninth European Conference on Computer-Supported Cooperative Work.*

Novick, D. G., & Ward, K. (2006). Why don't people read the manual? *Proceedings of the 24th Annual Conference on Design of Communication—SIGDOC '06*, 11. https://doi.org/10.1145/1166324.1166329

Ojer, M., Alvarez, H., Serrano, I., Saiz, F. A., Barandiaran, I., Aguinaga, D., Querejeta, L., & Alejandro, D. (2020). Projection-based augmented reality assistance for manual electronic component assembly processes. *Applied Sciences Switzerland, 10*(3). https://doi.org/10.3390/app10030796

O'Neill, J., Castellani, S., Grasso, A., Roulland, F., & Tolmie, P. (2005). Representations can be good enough. In H. Gellersen, K. Schmidt, M. Beaudouin-Lafon, & W. Mackay (Eds.), *ECSCW 2005* (pp. 267–286). Springer Netherlands. https://doi.org/10.1007/1-4020-4023-7_14

Ong, S. K., Yuan, M. L., & Nee, A. Y. C. (2008). Augmented reality applications in manufacturing: A survey. *International Journal of Production Research, 46*(10), 2707–2742. https://doi.org/10.1080/00207540601064773

Orlikowski, W. J. (1993). Learning from notes: Organizational issues in groupware implementation. *Information Society, 9*(3), 237–250. https://doi.org/10.1080/01972243.1993.9960143

Orr, J. E. (1996). *Talking about machines: An ethnography of a modern job*. Cornell University Press.

Otto, J., Henning, S., & Niggemann, O. (2014). Why cyber-physical production systems need a descriptive engineering approach—A case study in plug & produce. *Procedia Technology, 15*, 295–302. https://doi.org/10.1016/j.protcy.2014.09.083

Paas, F., Renkl, A., & Sweller, J. (2003). Cognitive load theory and instructional design: Recent developments. *Educational Psychologist, 38*(1), 1–4. https://doi.org/10.1207/S15326985EP3801_1

Pacaux-Lemoine, M. P., Berdal, Q., Enjalbert, S., & Trentesaux, D. (2018). Towards human-based industrial cyber-physical systems. *Proceedings—2018 IEEE Industrial Cyber-Physical Systems, ICPS 2018*, 615–620. https://doi.org/10.1109/ICPHYS.2018.8390776

Paelke, V., & Röcker, C. (2015). User interfaces for cyber-physical systems: Challenges and possible approaches. In A. Marcus (Ed.), *Design, user experience, and usability: Design discourse: 4th international conference, DUXU 2015* (Vol. 9186). Springer International Publishing. https://doi.org/10.1007/978-3-319-20886-2

Palanisamy, S., & Siddiqui, S. (2007). Changeover time reduction and productivity improvement by integrating conventional SMED method with implementation of MES for better production planning and control. *International Journal of Innovative Research in Science, Engineering and Technology (An ISO Certified Organization), 3297*(12), 7961–7974.

Park, K.-J., Zheng, R., & Liu, X. (2012). Cyber-physical systems: Milestones and research challenges. *Journal Computer Communications, 36*(1), 1–7.

Park, M., Schmidt, L., Schlick, C., & Luczak, H. (2007). Design and evaluation of an augmented reality welding helmet. *Human Factors and Ergonomics in Manufacturing, 17*(4), 317–330. https://doi.org/10.1002/hfm

Pathomaree, N., & Charoenseang, S. (2005). Augmented reality for skill transfer in assembly tasks. *ROMAN 2005. IEEE International Workshop on Robot and Human Interactive Communication, 2005*, 500–504. https://doi.org/10.1109/ROMAN.2005.1513829

Patterson, R. E., Blaha, L. M., Grinstein, G. G., Liggett, K. K., Kaveney, D. E., Sheldon, K. C., Havig, P. R., & Moore, J. A. (2014). A human cognition framework for information visualization. *Computers and Graphics (Pergamon), 42*(1), 42–58. https://doi.org/10.1016/j.cag.2014.03.002

Peinl, R. (2006). A knowledge sharing model illustrated with the software development industry. In F. Lehner, H. Nösekabel, & P. Kleinschmidt (Eds.), *Multikonferenz Wirtschaftsinformatik 2006* (Vol. 2, pp. 389–401). GITO-Verlag.

Peruzzini, M., & Pellicciari, M. (2017). A framework to design a human-centred adaptive manufacturing system for aging workers. *Advanced Engineering Informatics, 33*, 330–349. https://doi.org/10.1016/j.aei.2017.02.003

Petersen, N., Pagani, A., & Stricker, D. (2013). Real-time modeling and tracking manual workflows from first-person vision. *2013 IEEE International Symposium on Mixed and Augmented Reality, ISMAR 2013*, 117–124. https://doi.org/10.1109/ISMAR.2013.6671771

Pipek, V., Wulf, V., & Johri, A. (2012). Bridging artifacts and actors: Expertise sharing in organizational ecosystems. *Computer Supported Cooperative Work, 21*(2–3), 261–282. https://doi.org/10.1007/s10606-011-9138-y

Pisano, G. P. (1996). Learning-before-doing in the development of new process technology. *Research Policy*. https://doi.org/10.1016/S0048-7333(96)00896-7

Polanyi, M. (1967). Sense-giving and sense-reading. *Philosophy, 42*(162), 301–325.
Pollalis, C., Fahnbulleh, W., Tynes, J., & Shaer, O. (2017). HoloMuse. *Proceedings of the Tenth International Conference on Tangible, Embedded, and Embodied Interaction—TEI '17*, 565–570. https://doi.org/10.1145/3024969.3025094
Poltrock, S., & Grudin, J. (1997). CSCW, groupware and workflow: Experiences, state of art, and future trends. In S. Howard, J. Hammond, & G. Lindgaard (Eds.), *Human-Computer Interaction—Interact '97* (pp. 661–662). Chapman & Hall.
Ponton, M. K., Gail Derrick, M., & Carr, P. B. (2005). The relationship between resourcefulness and persistence in adult autonomous learning. *Adult Education Quarterly, 55*(2), 116–128. https://doi.org/10.1177/0741713604271848
Poovendran, R., Sampigethaya, K., Gupta, S. K. S., Lee, I., Prasad, K. V., Corman, D., & Paunicka, J. L. (2012). Special issue on cyber-physical systems. *Proceedings of the IEEE, 100*(1), 6–12. https://doi.org/10.1109/jproc.2011.2167449
Post, T., Ilsen, R., Hamann, B., Hagen, H., & Aurich, J. C. (2017). User-guided visual analysis of cyber-physical production systems. *Journal of Computing and Information Science in Engineering, 17*(2). https://doi.org/10.1115/1.4034872
Radkowski, R., Herrema, J., & Oliver, J. (2015). Augmented reality-based manual assembly support with visual features for different degrees of difficulty. *International Journal of Human-Computer Interaction, 31*(5), 337–349. https://doi.org/10.1080/10447318.2014.994194
Raelin, J. A. (1997). A model of work-based learning. *Organization Science, 8*(6), 563–578.
Rajkumar, R. (2012). A cyber physical future. *Proceedings of the IEEE, 100*(Special Centennial Issue), 1309–1312. https://doi.org/10.1109/JPROC.2012.2189915
Rajkumar, R., Lee, I. L., Sha, L. S., & Stankovic, J. (2010). Cyber-physical systems: The next computing revolution. *Design Automation Conference (DAC), 2010 47th ACM/IEEE*, 0–5. https://doi.org/10.1145/1837274.1837461
Ramakers, R., Anderson, F., Grossman, T., & Fitzmaurice, G. (2016). RetroFab: A design tool for retrofitting physical interfaces using actuators, sensors and 3D printing. *Conference on Human Factors in Computing Systems—Proceedings*, 409–419. https://doi.org/10.1145/2858036.2858485
Ramirez, H., Mendivil, E. G., Flores, P. R., & Gonzalez, M. C. (2013). Authoring software for augmented reality applications for the use of maintenance and training process. *Procedia Computer Science, 25*, 189–193.
Randall, D., Brien, J. O., Rouncefield, M., & Hughes, J. A. (1996). Organisational memory and CSCW: Supporting the "Mavis Phenomenon." *Proceedings Sixth Australian Conference on Computer-Human Interaction*, 26–33. https://doi.org/10.1109/OZCHI.1996.559984
Randall, D., Hughes, J. A., O'Brien, J., Rodden, T., Rouncefield, M., Sommerville, I., & Tolmie, P. (1998). Banking on the old technology: Understanding the organizational context of 'legacy' issues. Presented at the *Labour Process Conference, November*.
Reichling, T. (2008). Wissensmanagement in einer Netzwerkorganisation. Wulf, Volker ua. (Hrsg.): *Reihe Schriften zu Kooperations- und Mediensystemen*. Bd, 21.
Reuter, C., Ludwig, T., & Mischur, P. (2019). RescueGlass: Collaborative applications involving head-mounted displays for Red Cross rescue dog units. *Computer Supported Cooperative Work (CSCW), 28*(1–2), 209–246. https://doi.org/10.1007/s10606-018-9339-8

Roberto, R. A., Lima, J. P., Mota, R. C., & Teichrieb, V. (2016). Authoring tools for augmented reality: An analysis and classification of content design tools. *International Conference of Design, User Experience, and Usability, 9748*, 237–248. https://doi.org/10.1007/978-3-319-40406-6_22

Roberts, J. (2000). From know-how to show-how? Questioning the role of information and communication technologies in knowledge transfer. *Technology Analysis & Strategic Management, 12*(4), 429–443. https://doi.org/10.1080/713698499

Robertson, T., & Wagner, I. (2015). CSCW and the Internet of Things. September 2015, 19–23. https://doi.org/10.1007/978-3-319-20499-4

Robinson, M., Kovalainen, M., & Auramäki, E. (2000). Diary as dialogue in papermill process control. *Communications of the ACM, 43*(1), 65–70. https://doi.org/10.1145/323830.323846

Rodriguez, L., Quint, F., Gorecky, D., Romero, D., & Siller, H. R. (2015). Developing a mixed reality assistance system based on projection mapping technology for manual operations at assembly workstations. *Procedia Computer Science, 75*(Vare), 327–333. https://doi.org/10.1016/j.procs.2015.12.254

Rohde, M., Brödner, P., Stevens, G., Betz, M., & Wulf, V. (2017). Grounded design—A praxeological IS research perspective. *Journal of Information Technology, 32*(2), 163–179. https://doi.org/10.1057/jit.2016.5

Romero, D., Bernus, P., Noran, O., Stahre, J., & Berglund, Å. F. (2016). The operator 4.0: Human cyber-physical systems & adaptive automation towards human-automation symbiosis work systems. *IFIP Advances in Information and Communication Technology, 488*, 677–686. https://doi.org/10.1007/978-3-319-51133-7_80

Ryle, G. (1945). Knowing how and knowing that: The presidential address. *Proceedings of the Aristotelian Society, 46*(1945–1946), 1–16.

Ryle, G. (2009). *The concept of mind*. Routledge. https://doi.org/10.4324/9780203875858

Rzayev, R., Woźniak, P. W., Dingler, T., & Henze, N. (2018). Reading on smart glasses: The effect of text position, presentation type and walking. *Proceedings of the 2018 CHI Conference on Human Factors in Computing Systems*. https://doi.org/10.1145/3173574.3173619

Sacks, H. (1995). Lectures on Conversation. Cambridge, MA, USA: Blackwell Publishers.

Salvador, T., & Mateas, M. (1997). Introduction to design ethnography. *Conference on Human Factors in Computing Systems—Proceedings, 22–27-March-1997*, 166–167. https://doi.org/10.1145/1120212.1120325

Sanna, A., Manuri, F., Lamberti, F., Paravati, G., & Pezzolla, P. (2015). Using handheld devices to support augmented reality-based maintenance and assembly tasks. *2015 IEEE International Conference on Consumer Electronics (ICCE)*, 178–179. https://doi.org/10.1109/ICCE.2015.7066370

Schatzki, T. R. (2001). *The practice turn in contemporary theory*. Routledge.

Schirner, G., Erdogmus, D., Chowdhury, K., & Padir, T. (2013). The future of human-in-the-loop cyber-physical systems. *Computer, 46*(1), 36–45. https://doi.org/10.1109/MC.2013.31

Schlechtendahl, J., Keinert, M., Kretschmer, F., Lechler, A., & Verl, A. (2015). Making existing production systems Industry 4.0-ready: Holistic approach to the integration of existing production systems in Industry 4.0 environments. *Production Engineering, 9*(1), 143–148. https://doi.org/10.1007/S11740-014-0586-3

Schlegel, A., Wenzel, K., & Hoffmann, M. (2018). *SMARTE FABRIK—Digitalisierung und Automatisierung*. Fraunhofer IWU.

Schmidt, K. (1991). Computer support for cooperative work in advanced manufacturing. *The International Journal of Human Factors in Manufacturing, 1*(4), 303–320. https://doi.org/10.1002/hfm.4530010402

Schmidt, K. (2012). The trouble with "tacit knowledge." *Computer Supported Cooperative Work, 21*(2–3), 163–225. https://doi.org/10.1007/s10606-012-9160-8

Schmidt, K. (2014). The concept of "practice": What's the point? In C. Rossitto, L. Ciolfi, D. Martin, & B. Conein (Eds.), *COOP 2014. Proceedings of the 11th International Conference on the Design of Cooperative Systems* (pp. 427–444). Springer.

Schmidt, K., & Bannon, L. (1992). Taking CSCW seriously: Supporting articulation work. *Computer Supported Cooperative Work, 1*(1–2), 7–40. https://doi.org/10.1007/978-1-84800-068-1_3

Schnotz, W., & Bannert, M. (2003). Construction and interference in learning from multiple representations. *Learning and Instruction, 13*(2), 141–156. https://doi.org/10.1016/S0959-4752(02)00017-8

Schrepp, M., Hinderks, A., & Thomaschewski, J. (2014). Applying the User Experience Questionnaire (UEQ) in different evaluation scenarios. In *Lecture Notes in Computer Science (including subseries Lecture Notes in Artificial Intelligence and Lecture Notes in Bioinformatics)*: Vol. 8517 LNCS (Issue PART 1, pp. 383–392). https://doi.org/10.1007/978-3-319-07668-3_37

Schuh, G., Gartzen, T., Rodenhauser, T., & Marks, A. (2015). Promoting work-based learning through industry 4.0. *Procedia CIRP, 32*(Clf), 82–87. https://doi.org/10.1016/j.procir.2015.02.213

Schuh, G., Potente, T., Wesch-Potente, C., Weber, A. R., & Prote, J. P. (2014). Collaboration mechanisms to increase productivity in the context of Industrie 4.0. *Procedia CIRP, 19*(C), 51–56. https://doi.org/10.1016/j.procir.2014.05.016

Schwab, K. (2017). *The fourth industrial revolution*. Crown Currency.

Seichter, H., Looser, J., & Billinghurst, M. (2008). ComposAR: An intuitive tool for authoring AR applications. *Proceedings—7th IEEE International Symposium on Mixed and Augmented Reality 2008, ISMAR 2008*, 177–178. https://doi.org/10.1109/ISMAR.2008.4637354

Shah, M. M., Arshad, H., & Sulaiman, R. (2012). Occlusion in augmented reality. In S. Fong, K. D. Kwack, & F. Ko (Eds.), *8th International Conference on Information Science and Digital Content Technology (ICIDT2012)* (pp. 372–378). IEEE.

Shapiro, S. S., & Wilk, M. B. (1965). An analysis of variance test for normality (complete samples). *Biometrika, 52*(3/4), 591. https://doi.org/10.2307/2333709

Sharp, H., Rogers, Y., & Preece, J. (2006). *Interaction design: Beyond human-computer interaction* (2nd ed.). John Wiley & Sons.

Shi, J., Wan, J., Yan, H., & Suo, H. (2011). A survey of cyber-physical systems. *Wireless Communications and Signal Processing (WCSP)*, 1–6.

Shingo, S. (1985). *A revolution in manufacturing: The SMED system*. CRC Press.

Shu, Q., Tu, Q., & Wang, K. (2011). The impact of computer self-efficacy and technology dependence on computer-related technostress: A social cognitive theory perspective. *International Journal of Human-Computer Interaction, 27*(10), 923–939. https://doi.org/10.1080/10447318.2011.555313

Siltanen, S., Hakkarainen, M., Korkalo, O., Salonen, T., Sääski, J., Woodward, C., Kannetis, T., Perakakis, M., & Potamianos, A. (2007). Multimodal user interface for augmented assembly. *2007 IEEE 9Th International Workshop on Multimedia Signal Processing, MMSP 2007—Proceedings*, 78–81. https://doi.org/10.1109/MMSP.2007.4412822

Simon, R., Sclabassi, R., & Znati, T. (1994). Communication control in computer supported cooperative work systems. *Proceedings of the 1994 ACM Conference on Computer Supported Cooperative Work, CSCW 1994*, 311–321. https://doi.org/10.1145/192844.193036

Smith, E., Semple, G., Evans, D., McRae, K., & Blackwell, P. (2020). Augmented instructions: Analysis of performance and efficiency of assembly tasks. In J. Y. C. Chen & G. Fragomeni (Eds.), *Virtual, augmented and mixed reality. Industrial and everyday life applications* (Vol. 12191, pp. 166–177). Springer International Publishing. https://doi.org/10.1007/978-3-030-49698-2_12

Speier, C., Valacich, J. S., & Vessey, I. (1997). The effects of task interruption and information presentation on individual decision making. *Proceedings of the 18th International Conference on Information Systems, ICIS 1997, 30*(2), 21–35.

Speier, C., Vessey, I., & Valacich, J. S. (2003). The effects of interruptions, task complexity, and information presentation on computer-supported decision-making performance. *Decision Sciences, 34*(4), 771–797. https://doi.org/10.1111/j.1540-5414.2003.02292.x

Spender, J.-C. (1996a). Organizational knowledge, learning and memory: Three concepts in search of a theory. *Journal of Organizational Change Management, 9*(1), 63–78. https://doi.org/10.1108/09534819610156813

Spender, J.-C. (1996b). Making knowledge the basis of a dynamic theory of the firm. *Strategic Management Journal, 17*(S2), 45–62. https://doi.org/10.1002/smj.4250171106

Stevens, G., Pipek, V., & Wulf, V. (2009). Appropriation infrastructure: Supporting the design of usages. *End-User Development. IS-EUD 2009. Lecture Notes in Computer Science, 5435 LNCS*, 50–69. https://doi.org/10.1007/978-3-642-00427-8_4

Stevens, G., Rohde, M., Korn, M., & Wulf, V. (2018). Grounded design: A research paradigm in practice-based computing. In V. Wulf, V. Pipek, M. Rohde, G. Stevens, & D. Randall (Eds.), *Socio-Informatics: A practice-based perspective on the design and use of IT artifacts* (pp. 23–46). Oxford University Press.

Stone, E. F., & Stone, D. L. (1984). The effects of multiple sources of performance feedback and feedback favorability on self-perceived task competence and perceived feedback accuracy. *Journal of Management, 10*(3), 371–378. https://doi.org/10.1177/014920638401000311

Strauss, A., Fargerhaugh, S., Suczek, B., & Wiener, C. (1985). *Social organisation of medical work*. University of Chicago Press.

Strohbach, M., Gellersen, H., Kortuem, G., & Kray, C. (2004). Cooperative artefacts: Assessing real-world situations with embedded technology. *International Conference on Ubiquitous Computing*, 250–267.

Svensson, M. S., Heath, C., & Luff, P. (2007). Instrumental action: The timely exchange of implements during surgical operations. *ECSCW 2007*. https://doi.org/10.1007/978-1-84800-031-5_3

Sweller, J. (1994). Cognitive load theory, learning difficulty, and instructional design. *Learning and Instruction, 4*, 295–312. https://doi.org/10.1016/0959-4752(94)90003-5

Syberfeldt, A., Danielsson, O., Holm, M., & Wang, L. (2015). Visual assembling guidance using augmented reality. *Procedia Manufacturing, 1*, 98–109. https://doi.org/10.1016/j.promfg.2015.09.068

Szulanski, G. (2000). The process of knowledge transfer: A diachronic analysis of stickiness. *Organizational Behavior and Human Decision Processes, 81*(1), 9–27. https://doi.org/10.1006/obhd.2000.2884.

Tamhane, A. C. (1977). Multiple comparisons in model I one-way ANOVA with unequal variances. *Communications in Statistics—Theory and Methods, 6*(1), 15–32. https://doi.org/10.1080/03610927708827466

Tang, A., Neustaedter, C., & Greenberg, S. (2007). Videoarms: Embodiments for mixed presence groupware. In: Bryan-Kinns, N., Blanford, A., Curzon, P., Nigay, L. (eds): *People and Computers XX—Engage*. Springer, London. https://doi.org/10.1007/978-1-84628-664-3_8

Tang, A., Owen, C., Biocca, F., & Mou, W. (2003). Comparative effectiveness of augmented reality in object assembly. *Proceedings of the Conference on Human Factors in Computing Systems—CHI '03*, 9, 73. https://doi.org/10.1145/642625.642626

Tantik, E., & Anderl, R. (2016). Industrie 4.0: Using Cyber-physical Systems for Value-stream Based Production Evaluation. *Procedia CIRP, 57*, 207–212. https://doi.org/10.1016/j.procir.2016.11.036

Tausczik, Y. R., & Pennebaker, J. W. (2012). Participation in an online mathematics community: Differentiating motivations to add. *Organization, 207–216*. https://doi.org/10.1145/2145204.2145237

Thramboulidis, K., & Christoulakis, F. (2016). UML4IoT—A UML-based approach to exploit IoT in cyber-physical manufacturing systems. *Computers in Industry, 82*, 259–272. https://doi.org/10.1016/j.compind.2016.05.010

Tomarken, A. J., & Serlin, R. C. (1986). Comparison of ANOVA alternatives under variance heterogeneity and specific noncentrality structures. *Psychological Bulletin, 99*(1), 90–99. https://doi.org/10.1037/0033-2909.99.1.90

Treude, C., Barzilay, O., & Storey, M. A. (2011). How do programmers ask and answer questions on the web? *Proceedings—International Conference on Software Engineering*, 804–807. https://doi.org/10.1145/1985793.1985907

Tuomi, I. (1999). Data is more than knowledge: Implications of the reversed knowledge hierarchy for knowledge management and organizational memory. *Proceedings of the 32nd Annual Hawaii International Conference on Systems Sciences. 1999. HICSS-32. Abstracts and CD-ROM of Full Papers, Track1*(c), 12. https://doi.org/10.1109/HICSS.1999.772795

Turkle, S. (2002). Sociable technologies: Enhancing human performance. *Converging Technologies for Improving Human Performance*, 150.

Twidale, M., Randall, D., & Bentley, R. (1994). Situated evaluation for cooperative systems. *Proceedings of the 1994 ACM Conference on Computer Supported Cooperative Work, CSCW 1994*, 441–452. https://doi.org/10.1145/192844.193066

Ullah, I., & Narain, R. (2021). Achieving mass customization capability: The roles of flexible manufacturing competence and workforce management practices. *Journal of Advances in Management Research, 18*(2), 273–296. https://doi.org/10.1108/JAMR-05-2020-0067

Uva, A. E., Gattullo, M., Manghisi, V. M., Spagnulo, D., Cascella, G. L., & Fiorentino, M. (2018). Evaluating the effectiveness of spatial augmented reality in smart manufacturing:

A solution for manual working stations. *International Journal of Advanced Manufacturing Technology, 94*(1–4), 509–521. https://doi.org/10.1007/s00170-017-0846-4
Vargha, A., & Delaney, H. D. (1998). The Kruskal-Wallis test and stochastic homogeneity. *Journal of Educational and Behavioral Statistics, 23*(2), 170–192. https://doi.org/10.3102/10769986023002170
Vasilescu, B., Filkov, V., & Serebrenik, A. (2013). Stackoverflow and github: Associations between software development and crowdsourced knowledge. *2013 International Conference on Social Computing*, 188–195.
Vasilescu, B., Serebrenik, A., Devanbu, P., & Filkov, V. (2014). How social Q&A sites are changing knowledge sharing in open source software communities. *Proceedings of the ACM Conference on Computer Supported Cooperative Work, CSCW*, 342–354. https://doi.org/10.1145/2531602.2531659
VDI 3430. (2014). *Rotationszugbiegen von Profilen.*
Velden, T. (2013). Explaining field differences in openness and sharing in scientific communities. *Proceedings of the ACM Conference on Computer Supported Cooperative Work, CSCW*, 445–457. https://doi.org/10.1145/2441776.2441827
Vicente, K. J. (2000). HCI in the global knowledge-based economy: Designing to support worker adaptation. *ACM Transactions on Computer-Human Interaction, 7*(2), 263–280. https://doi.org/10.1145/353485.353489
Viljakainen, T. (2020). Adoption of augmented reality solutions in field engineering and maintenance: Drivers and barriers for organizations.
Voigt, S., Seidel, H., Orth, R., & Kohl, H. (2016). Herausforderung für Unternehmen. In *Wissensmanagement im Mittelstand* (pp. 9–18). https://doi.org/10.1007/978-3-662-49220-8_2
Wagner, R. M. (Ed.). (2018). *Industrie 4.0 für die Praxis.* Springer Gabler.
Wakkary, R., Poon, M., Maestri, L., Kirton, T., Julihn, C., & Betts, R. (2007). How informances can be used in design ethnography. *Conference on Human Factors in Computing Systems—Proceedings*, 1875–1880. https://doi.org/10.1145/1240866.1240914
Wan, J., Cai, H., & Zhou, K. (2015). Industrie 4.0: Enabling technologies. *Proceedings of 2015 International Conference on Intelligent Computing and Internet of Things*, 135–140. https://doi.org/10.1109/ICAIOT.2015.7111555
Wang, F. (2010). The emergence of intelligent enterprises: From CPS to CPSS. *IEEE Intelligent Systems, 25*(4), 85–88.
Wang, J., Feng, Y., Zeng, C., & Li, S. (2014). An augmented reality-based system for remote collaborative maintenance instruction of complex products. *IEEE International Conference on Automation Science and Engineering, 2014-Janua*, 309–314. https://doi.org/10.1109/CoASE.2014.6899343
Wang, L., Törngren, M., & Onori, M. (2015). Current status and advancement of cyber-physical systems in manufacturing. *Journal of Manufacturing Systems, 37*, 517–527. https://doi.org/10.1016/j.jmsy.2015.04.008
Wang, X., & Dunston, P. S. (2006). Compatibility issues in augmented reality systems for AEC: An experimental prototype study. *Automation in Construction, 15*(3), 314–326. https://doi.org/10.1016/j.autcon.2005.06.002
Wang, X., Ong, S. K., & Nee, A. Y. C. (2016). A comprehensive survey of augmented reality assembly research. *Advances in Manufacturing, 4*(1), 1–22. https://doi.org/10.1007/s40436-015-0131-4

Watson, S., & Hewett, K. (2006). A multi-theoretical model of knowledge transfer in organizations: Determinants of knowledge contribution and knowledge reuse. *Journal of Management Studies, 43*(2), 141–173. https://doi.org/10.1111/j.1467-6486.2006.00586.x

Wenger, E. (1998). Communities of practice: Learning as a social system. *Systems Thinker, 9*(5), 2–3. https://doi.org/10.2277/0521663636

Werrlich, S., Daniel, A., Ginger, A., Nguyen, P. A., & Notni, G. (2019). Comparing HMD-based and paper-based training. *Proceedings of the 2018 IEEE International Symposium on Mixed and Augmented Reality, ISMAR 2018*, 134–142. https://doi.org/10.1109/ISMAR.2018.00046

Werrlich, S., Lorber, C., Nguyen, P.-A., Yanez, C. E. F., & Notni, G. (2018). Assembly training: Comparing the effects of head-mounted displays and face-to-face training. In *Lecture Notes in Computer Science (including subseries Lecture Notes in Artificial Intelligence and Lecture Notes in Bioinformatics)* (pp. 462–476). Springer International Publishing. https://doi.org/10.1007/978-3-319-91581-4_35

Westerfield, G., Mitrovic, A., & Billinghurst, M. (2015). Intelligent augmented reality training for motherboard assembly. *International Journal of Artificial Intelligence in Education, 25*(1), 157–172. https://doi.org/10.1007/s40593-014-0032-x

Whalen, J. (1995). A technology of order production: Computer-aided dispatch in public safety communications. In *Situated Order: Studies in the Social Organization of Talk and Embodied Activities.*

Whalen, J., Whalen, M., & Henderson, K. (2002). Improvisational choreography in teleservice work. *British Journal of Sociology*. https://doi.org/10.1080/00071310220133322

Wiedenmaier, S., Oehme, O., Schmidt, L., & Luczak, H. (2003). Augmented reality (AR) for assembly processes: Design and experimental evaluation. *International Journal of Human-Computer Interaction, 16*(3), 497–514. https://doi.org/10.1207/S15327590IJHC1603_7

Wilschut, E. S., Murphy, M. S., & Bosch, T. (2019). Using chunking of instructions, spatial augmented reality and display-based work instructions. *Petra'19*, 376–381.

Winch, P. (1964). Understanding a primitive society. *American Philosophical Quarterly, 1*(4), 307–324.

Winch, P. (1997). Can we understand ourselves? *Philosophical Investigations, 20*(3), 193–204. https://doi.org/10.1111/1467-9205.00038

Wittgenstein, L. (1953). *Philosophical investigations.* Blackwell.

Wolf, W. (2009). Cyber-physical systems. *Computers, 42*(3), 88–89. https://doi.org/10.1109/MC.2009.81

Wollersheim, J., Leyer, M., & Spörrle, M. (2015). When more is not better: The effect of the number of learning interventions on the acquisition of process-oriented thinking. *Management Learning, 47*(2), 137–157. https://doi.org/10.1177/1350507615587447

Wright, P. (1977). Presenting technical information: A survey of research findings. *Instructional Science, 6*(2), 93–134. https://doi.org/10.1007/BF00121082

Wulf, V., Müller, C., Pipek, V., Randall, D., Rohde, M., & Stevens, G. (2015). Practice-based computing: Empirically grounded conceptualizations derived from design case studies. In *Springer London* (pp. 111–150). https://doi.org/10.1007/978-1-4471-6720-4_7

Wulf, V., Pipek, V., Randall, D., Rohde, M., Schmidt, K., & Stevens, G. (2018). *Socio-informatics: A practice-based perspective on the design and use of IT artifacts.* Oxford University Press.

Wulf, V., Rohde, M., Pipek, V., & Stevens, G. (2011). Engaging with practices: Design case studies as a research framework in CSCW. *Proceedings of the ACM 2011 Conference on Computer Supported Cooperative Work—CSCW '11*, 505–512. https://doi.org/10.1145/1958824.1958902

Yamauchi, Y., Whalen, J., & Bobrow, D. G. (2003). Information use of service technicians in difficult cases. *Proceedings of the SIGCHI Conference on Human Factors in Computing Systems*, 81–88. https://doi.org/10.1145/642611.642627

Yang, J., Adamic, L. A., & Ackerman, M. S. (2008). Crowdsourcing and knowledge sharing: Strategic user behavior on Taskcn. *Proceedings of the 9th ACM Conference on Electronic Commerce*, 246–255. https://doi.org/10.1145/1386790.1386829

Yuan, M. L., Ong, S. K., & Nee, A. Y. C. (2008). Augmented reality for assembly guidance using a virtual interactive tool. *International Journal of Production Research, 46*(7), 1745–1767. https://doi.org/10.1080/00207540600972935

Yuen, S. C.-Y., Yaoyuneyong, G., & Johnson, E. (2011). Augmented reality: An overview and five directions for AR in education. *Journal of Educational Technology Development and Exchange, 4*(1). https://doi.org/10.18785/jetde.0401.10

Zauner, J., Haller, M., Brandl, A., & Hartmann, W. (2003). Authoring of a mixed reality furniture assembly instructor for hierarchical structures. *Proceedings of the SIGGRAPH 2003 Conference on Sketches & Applications in Conjunction with the 30th Annual Conference on Computer Graphics and Interactive Techniques—GRAPH '03*, 1. https://doi.org/10.1145/965400.965448

Zheng, X. S., Foucault, C., Matos da Silva, P., Dasari, S., Yang, T., & Goose, S. (2015). Eyewearable technology for machine maintenance. *Proceedings of the 33rd Annual ACM Conference on Human Factors in Computing Systems—CHI '15*, 2125–2134. https://doi.org/10.1145/2702123.2702305

Zhou, K., Liu, T., & Zhou, L. (2015). Industry 4.0: Towards future industrial opportunities and challenges. *2015 12th International Conference on Fuzzy Systems and Knowledge Discovery (FSKD)*, 2147–2152. https://doi.org/10.1109/FSKD.2015.7382284

Zhu, J., Ong, S. K., & Nee, A. Y. C. (2014). A context-aware augmented reality system to assist the maintenance operators. *International Journal on Interactive Design and Manufacturing, 8*(4), 293–304. https://doi.org/10.1007/s12008-013-0199-7

Zijlstra, F. R. H., & Van Doorn, L. (1985). The construction of a scale to measure perceived effort. *University of Technology.*

Zijlstra, H. (1993). Efficiency in work behavior. A design approach for modern tools. *Delft University of Technology.*

MIX
Papier aus verantwortungsvollen Quellen
Paper from responsible sources
FSC® C105338

If you have any concerns about our products,
you can contact us on
ProductSafety@springernature.com

In case Publisher is established outside the EU,
the EU authorized representative is:
Springer Nature Customer Service Center GmbH
Europaplatz 3, 69115 Heidelberg, Germany

Printed by Libri Plureos GmbH
in Hamburg, Germany